Canonical Forms in Finitely Presented Algebras

Philippe Le Chenadec
INRIA, France

Canonical Forms in Finitely Presented Algebras

Pitman, London

John Wiley & Sons, Inc., New York, Toronto

PITMAN PUBLISHING LIMITED
128 Long Acre, London WC2E 9AN

A Longman Group Company

First published 1986

Available in the Western Hemisphere from
John Wiley & Sons, Inc.
605 Third Avenue, New York, NY 10158

British Library Cataloguing in Publication Data

Le Chenadec, Philippe
Canonical forms in finitely presented algebras.
—(Research notes in theoretical computer science, ISSN 0268-7534)
1. Equations 2. Forms (Mathematics)
I. Title II. Series
512.9'4 QA214

ISBN 0-273-08721-5 (Pitman)

Library of Congress Cataloging in Publication Data

Le Chenadec, Philippe.
Canonical forms in finitely presented algebras.

Bibliography: p.
1. Algebra, Universal. I. Title.
QA251.L4 1986 512 85-30105
ISBN 0-470-20291-2 (Wiley)

Reproduced and printed by photolithography
in Great Britain by Biddles Ltd, Guildford

Foreword

Research in theoretical computer science has experienced tremendous growth both in the depth to which older theories have been pursued and also in the number of new problem areas that have arisen. While theoretical computer science is mathematical in nature, its goals include the development of an understanding of the nature of computation as well as the solution of specific problems that arise in the practice of computing.

The purpose of this series of monographs is to make available to the professional community expositions of topics that play an important role in theoretical computer science or that provide bridges with other aspects of computer science and with aspects of mathematics. The scope of the series may be considered to be that represented by the leading journals in the field. The editors intend that the scope will expand as the field grows and welcome submissions from all of those interested in theoretical computer science.

Ronald V. Book
Main Editor

Contents

Introduction

At least once, every student in mathematics has been confused while proving an equality in some algebraic structure by means of elementary operations, i.e. substitutions and replacements of equals by equals. This monograph is an introduction to rewriting systems of algebras defined by a finite set of identities. Rewriting systems, developed in the last two decades, attempts to solve this equality problem via computation of normal forms. This notion of rewriting or simplification has emerged as a crucial component in software performing symbolic computations.

We may classify such symbolic systems in two major trends. One tries to cover the essential algebraic areas by a collection of programs solving precise problems, polynomial factorization, integration, differentation, formal series summations, etc... Here, the most remarkable realizations are the Macsyma and Scratchpad systems [Fat79, Mac77, Bal85, Jen74, Gri71]. Other ones, such as Camac [Leo79] for finite groups, are more specialized (cf. [Buc82b] for a recent survey). This view of formal calculus is, of course, almost algorithmic, especially with respect to complexity. In its major performances, it is a powerful tool in proving some deep results, such as the enumeration of sporadic finite simple groups [Hal73, Leo77].

The other trend is based upon a precise formal definition of the manipulated objects. It appears historically with the works of Herbrand [Her30], Gentzen [Gen69] for the logical part; and in Robinson's work [Rob65] on automatic deduction by unification and resolution, or in recursive arithmetic allowing induction in the proofs of primitive recursive programs [Boy79]. But this orientation comes up against great difficulties due to inherent limitations: incompleteness results of Gödel [in Dav65], complexity results of Cook [Coo71], or some computing problems on proof structure [Goa80], computations in various specialized theories merged together [Nel79]. Also, most of the current systems are in fact interactive, proof control being assumed by the user (cf. [Ble84] for a recent overview of this research area). We mention the systems of Boyer and Moore [Boy79], AFFIRM [Mus80], LCF [Gor79] and Automath [Bru80] as representatives of such theorem-provers or proof-checkers.

Both tendencies include some *simplification* procedures in order to simplify the complex formulas generated in the course of a symbolic computation. This need for simplification has emerged as a new research domain, based on the formalization of these *reductions* in terms of computations on the axioms of an equational theory. This way was opened by a paper by D.E. Knuth and P.B. Bendix [Knu70], generalizing ideas present in T. Evan's analysis of normal forms [Eva51b].

With further investigations by G.E. Peterson and M.E. Stickel, D.S. Lankford and A.M. Ballantyne, G. Huet and J.M. Hullot [Pet81, Lan77a,b,c, Hul80b], a collection of complete rewriting systems for usual classes of algebras was found (groups, rings, distributive lattices...). These results enable an analysis of these structures under the rewriting approach. Such a study is the first goal of the present monograph. It shows that we may associate with many equational classes of algebras a compilation process of a *finitely presented algebra* A in the class. It translates a presentation of A into a complete set of *syntactic replacement rules*. This completeness states that the set of irreducible terms is a domain for a syntactic model of the algebra A. To these compilation processes (which possibly diverge), we give the generic name of *completion procedures*. It should be noted that the various completion procedures presented in chapters 2,3 and 4 are instances of the general Knuth-Bendix procedure through slight modifications. Hence, the correctness theorems and corollaries on termination are consequences of the general correctness theorem. Consequently, a complete proof of this last result is given, which is essentially Huet's one [Hue81].

Many complexity results claim the non-existence of efficient algorithms to solve a general problem. However, efficient algorithms working on a special kind of data frequently exist. And some of them are just translations of a general-purpose algorithm into a special-purpose one. The completion algorithms illustrate this fact: the most general one presented in chapter 2 is highly inefficient for our purposes, while its specialization through data and control structures gives tractable ones. For insights into these general/special-purpose algorithms, see the work of C. Goad [Goa80]. Also, this monograph gives some hints on what should be the unification of the above two major trends in Computing Algebra. Perhaps some evidence for this claim appears in the table of contents: the presentation starts with a survey of rewriting theory, a good representative of the second approach; while the following chapters enumerate various algorithms, encoding more and more knowledge about their inputs. Such an enumeration is similar to a presentation of formal systems of the first kind. Moreover, some algorithms informally described in the last chapter are efficient and special purpose ones. They solve linearly the word problem for some classes of geometric groups.

As far as algebra is concerned, the present work formalizes the notion of elementary computations in an algebra, given both the definition of its class (group or ring...) and its presentation by generators and relations. In so doing, we obtain a completion procedure of a finitely presented algebra, thus solving its word problem in the case of successful completion. Of course, in this survey of elementary algebra, we meet some well-known facts, they are considered from a constructive point of view, which after all is the one of computer science.

In semigroups or monoids, rewriting rules are called Thue or semi-Thue systems. They have been introduced by Thue and studied mainly for solvability results [Pos47, Boo82a, Ott84a] and their relation with automata theory and languages [Hop79, Dav58]. Abelian groups are shown to be direct products of cyclic groups. For other groups, we get two algorithms: the first one is the symmetrization of a group presentation and lies in the heart of small cancellation theory. Chapter 5 is an analysis of small cancellation from the rewriting point of view. This research was initiated by H. Bücken [Büc79a]. The specialist will find new diagrams characterizing small cancellation, rather different from usual ones. The second algorithm for groups is the usual completion. Many classical groups' families fall fairly under the scope of the algorithm, despite the unsolvability of the word problem in groups. Chapter 6 is a catalogue of complete systems for groups. Some nice interpretations of these systems on Cayley graphs show that completion provides a good means of description for these graphs. In ring theory, the rules rewrite a monomial into a polynomial. Two authors have investigated such systems: Bruno Buchberger [Buc65] has shown, for commutative polynomials over fields, the practical interest of these rules to solve various problems in commutative algebra. It is worth noting that the notion of a complete set of rules appears implicitly in Hironaka's famous paper [Hir64] where they are called standard basis. George M. Bergman's paper [Ber78] on the diamond lemma in rings is concerned with non-commutative rings. He outlines in some classical proofs the underlying presence of such rewriting systems. Completion may also be used for modules and algebras over a ring. This is allowed by the present approach where the rules'left members may include a coefficient from the scalar ring.

The author is indebted to Pierre-Louis Curien and the referee for their helpful comments on a draft of this work and to Gérard Huet who initiated this research.

1 Equational varieties

This first chapter presents some elementary notions of universal algebra. For a complete development including proofs we refer to the textbooks of Cohn or Grätzer [Coh65, Grä79]. Besides usual definitions, we introduce more technical ones, well-suited for the presentation of algorithms. The background of this chapter may be divided into three parts: pure algebra, abstract properties of reduction relations on a set and a detailed description of algebraic expressions in view of computer handling. We also introduce regular expressions used in the last chapter. An elementary background of recursion theory is assumed, see e.g. [Sho67, Hop79].

1.1 ALGEBRAS ON AN OPERATOR DOMAIN

First, let us fix some notations about relations over a set S. A binary relation R is a subset of the Cartesian product $S \times S$. Given two elements x and y in S, we write xRy when the pair (x,y) belongs to R. Let x,y,z be elements of S, we say that the relation R is:

reflexive	iff xRx.
irreflexive	iff one never has xRx.
symmetric	iff $xRy \Rightarrow yRx$.
antisymmetric	iff xRy and $yRx \Rightarrow x=y$.
transitive	iff xRy and $yRz \Rightarrow xRz$.
total	iff either xRy or yRx.
well-founded	iff there exists no infinite sequence $(x_i)_{i \in \mathbb{N}}$ of elements x_i in S such that for all i in $\mathbb{N}$, $x_i R x_{i+1}$.

This last property will also be called nœtherianity. An *equivalence relation* R is reflexive, symmetric and transitive. In this case, we may define the quotient set S/R as the set of equivalence classes.

A *partial-ordering* is a transitive and irreflexive relation. Adding the diagonal set $\Delta(S)=\{(x,x) \mid x \in S\}$ to a partial-ordering defines a *partial-order*, a transitive and reflexive relation. Conversely, the *strictpart* of a partial-order R, defined as $R-\{(x,x) \mid (x,x) \in R\}$, is a partial-ordering. A *quasi-order* is an antisymmetric partial-order. We will sometimes speak of an *ordering* abbreviating a partial-ordering. Given two relations R_1 and R_2, we define:

$R_1.R_2 = \{(x,y) \mid \exists z \in S\ xR_1z \text{ and } zR_2y\}$, the product of the two relations.

$R_1 \cup R_2 = \{(x,y) \mid xR_1y \text{ or } xR_2y\}$, their union.

$R_1^{\varepsilon} = R_1 \cup \Delta(S)$, the reflexive closure of R_1.

$R_1^{+} = \{(x,y) \mid \exists n>0,\ x=x_0, \ldots, x_n=y,\ x_i R_1 x_{i+1},\ 0 \leq i<n\}$, the transitive closure of R_1.

$R_1^{*} = R_1^{+} \cup \Delta(S)$, the transitive-reflexive closure of R_1.

To conclude these preliminaries, let us recall the principle of noetherian induction [Coh65,p.20]:

Let P be a predicate and $\rightarrow$ a noetherian relation on the set S,

if $\forall x \in S, (\forall y \in S$ with $x \stackrel{+}{\rightarrow} y,\ P(y))$, then $\forall x \in S,\ P(x)$.

Let F be a denumerable set whose elements are called operators or function symbols. To an operator **f** in the set F is associated a natural number $\alpha(\mathbf{f})$, the arity of **f**. The set F is called an operator domain. A constant is an operator with null arity. Operators will be denoted by **f**,**g**...,**+**,*****,... Let A be a set. To each operator **f**, we may associate a function $f^A : A^{\alpha(\mathbf{f})} \rightarrow A$.

Definition 1.1

A F-algebra is a pair $\mathbf{A}=(A,F^A)$, *where A is a non-empty set and* F^A *a set of functions* f^A *associated to each operator* **f** *in F.*

When clear from the context, the superscript A will be omitted and the F-algebra **A** simply called algebra. The set A is the carrier or the domain of the algebra. For example, if the operator domain F is defined by the two operators + and $-$ with $\alpha(+)=2$, $\alpha(-)=1$, then the set of integers modulo a natural number p is the carrier of a $\{+,-\}$-algebra, + and $-$ being the usual operations on integers modulo p. Another important example is the set of finite sequences over an arbitrary set E, noted E^*, defining a $\{.,\varepsilon\}$-algebra with two operators ., the concatenation of sequences, and ε, the empty sequence.

Let $\mathbf{B}=(B,F^B)$ be a F-algebra. Then, the algebra **B** is a subalgebra of **A** iff $B \subset A$ and, for all **f** in F, the restriction of f^A to the set B is equal to f^B, which is noted $f^A{\restriction}_B = f^B$.

Let $(\mathbf{A}_i)_I$ be a class of algebras indexed by a set I. A direct product of algebras is defined on the Cartesian product $A=\prod_I A_i$ by $\mathbf{A}=(A,F^A)$, the function $f^A \in F^A$ being defined by $f^A(a_1, \ldots, a_{\alpha(\mathbf{f})}) = (f^{A_i}(a_1^i, \ldots, a_{\alpha(\mathbf{f})}^i))_I$ where a_j^i is the ith-component of a_j in A. The algebra **A** is noted $\prod_I \mathbf{A}_i$.

In the same way, we define morphisms of algebras. If **A** and **B** are two F-algebras, then a function $\varphi : A \longrightarrow B$ is a morphism iff, for all **f** in F, we have

$$\varphi(f^A(a_1, \ldots, a_{\alpha(f)})) = f^B(\varphi(a_1), \ldots, \varphi(a_{\alpha(f)})).$$

Finally, a congruence on the algebra **A** is an equivalence relation $\sim$ compatible with the operators in F:

$$\forall f \in F,\ a_i \sim b_i\ i=1,\ldots,\alpha(f) \Rightarrow f^A(a_1, \ldots, a_{\alpha(f)}) \sim f^A(b_1, \ldots, b_{\alpha(f)}).$$

If the relation $\sim$ is a congruence on **A**, then the quotient set $A/\sim$ defines a F-algebra **A**$/\sim$, the quotient functions $f^A/\sim$ being the functions $f^{A/\sim}$. Finally, the image of the algebra **A** under a morphism is a subalgebra of **B** and is called a homomorphic image of **A**.

1.2 FREE ALGEBRAS AND THE TERMS

Now, we define the free F-algebra on a set V, whose elements will be later considered either as variables, constants or both, depending on the context. We also present the basic notion of **terms** inductively built on the operator domain F together with the variables (or constants) V. Informally, variables are quantified over all terms and they may be instanciated by any term. Constants may be added to the operator domain in order to distinguish special generating elements of an algebra.

Definition 1.2

Let **C** *be a class of F-algebras and V be a denumerable set disjoint from F. A F-algebra* **A**$=(A,F^A)$ *in* **C** *is* **C**-*free on* V *iff*

i) $V \subset A$,

ii) *For all F-algebras* **B**$=(B,F^B)$, *and every function* $\varphi : V \longrightarrow B$, *there exists a unique morphism* φ^* : **A**$\longrightarrow$**B**, *such that* $\varphi^*_{|V} = \varphi$.

Two **C**-free algebras on isomorphic sets are isomorphic. A theorem due to Birkhoff [Bir35] gives sufficient conditions for the existence of free algebras in a class:

Theorem 1.3 (Birkhoff)

Let C *be a class of F-algebras such that subalgebras, products and homomorphic images of F-algebras in* C *belong to* C, *then for all sets V, there exists a* C*-free algebra on V, and every algebra in* C *is the homomorphic image of a free* C*-algebra.*

Therefore, we can take the class Γ of all F-algebras for a given F. The previous theorem allows us to speak of the free F-algebra on V for short. This algebra is in fact the set of all expressions, or **first order terms,** built on F.

Definition 1.4

The F-algebra $\mathbf{T}(F,V) = (T,F)$ *where* T *is the subset of* $(F \cup V)^*$ *inductively defined by :*

i) $V \subset T$,

ii) $\forall\, \mathbf{f} \in F, \forall\, M_1, \ldots, M_{\alpha(\mathbf{f})} \in T,\ \mathbf{f}\, M_1 \cdots M_{\alpha(\mathbf{f})} \in T$.

is a free F-algebra on V.

The function f^T associates to the terms M_i the element $\mathbf{f} M_1 \cdots M_{\alpha(\mathbf{f})}$ in T. When V is empty, the free algebra is called the initial algebra, or the ground term algebra, a ground term being variable free. Note that the initial algebra is non-empty iff there exists at least one constant in F. Terms will be referred to by upper-case latin letters $M,N\ldots$, letters $a,b,c\ldots$ being reserved for constants and $x,y,z\ldots$ for variables. For the sake of clarity, we use infix notation and parenthesis for function operators. Let us list some technical definitions.

— The set of variables $V(M)$ of a term M.

i) $V(x) = \{x\}$ if $x \in V$, $V(c) = \phi$ if $c \in F$, $\alpha(c)=0$,

ii) $V(\mathbf{f}\, M_1 \cdots M_{\alpha(\mathbf{f})}) = \bigcup_{i=1}^{\alpha(\mathbf{f})} V(M_i)$.

— The set $O(M)$ of occurrences of the term M : an occurrence is an element of the free monoid $(\mathbb{N}^* - \{0\}, ., \varepsilon)$, with the concatenation and the empty occurrence ε.

i) $O(x) = \{\varepsilon\}$ if $x \in V$, $O(c) = \{\varepsilon\}$ if $c \in F$ and $\alpha(c) = 0$,

ii) $O(\mathbf{f} M_1 \cdots M_n) = \{\varepsilon\} \cup \{\, i.u \,/\, i=1,\ldots,n,\ u \in O(M_i)\}$.

Therefore, we can speak of the subterm of M at occurrence $u \in O(M)$, denoted by M/u :

i) $M/\varepsilon = M$,

ii) if $M = \mathbf{f}M_1 \cdots M_n$ then $M/iu = M_i/u$.

The term obtained from M by substituting the subterm M/u by the term N is noted $M[u \leftarrow N]$.

– Substitutions, noted $\sigma, \tau, \ldots$, are endomorphisms of $\mathbf{T}(F,V)$, defined by their values on V, extended to $\mathbf{T}(F,V)$ by congruence (cf. Def.1.2), and such that the domain $D(\sigma) = \{ x \mid x \in V \ \& \ \sigma(x) \neq x \}$ of the substitution σ is a finite set of variables. Thus, a substitution is defined by a finite set of pairs $\{<x,\sigma(x)> \mid x \in D(\sigma)\}$, and by the equalities $\sigma(\mathbf{f}\, M_1 \cdots M_n) = \mathbf{f}\, \sigma(M_1) ... \sigma(M_n)$. The union of τ and σ when $D(\sigma) \cap D(\tau) = \phi$ is denoted by $\sigma.\tau$, and more briefly $\sigma.<x,M>$ when $D(\tau) = \{x\}$. The composition of substitutions σ_1 and σ_2 is noted $\sigma_1 \circ \sigma_2$, $(\sigma_1 \circ \sigma_2)(M) = \sigma_1(\sigma_2(M))$.

Now, $\mathbf{T}(F,V)$ is partially ordered by the substitutions: $M \geq N$ iff there exists a substitution σ such that $\sigma(N) = M$. This relation is sometimes called the subsumption order of the term algebra [Hue80a]. The strict part of the subsumption ordering is well-founded [Hue76]. Two terms are $\geq$-equivalent iff they are equal modulo a bijection between $V(M)$ and $V(N)$. This congruence $=_\alpha$ is called the α-conversion of terms. It is easy to see that the substitution σ restricted to the set of variables of the term N (i.e. $\sigma(x)=x$ if $x \notin V(N)$) is unique modulo α-conversion. This restricted substitution is called a match from N to M. We can merge this ordering with the notion of subterms and define :

$$M \gg N \text{ iff } \exists u \in O(M), \sigma,\ M/u = \sigma(N) \text{ and } M \neq_\alpha N.$$

The ordering $\gg$ is also well-founded as composition of two noetherian orderings.

Observe that an efficient implementation of terms has to solve the α-conversion problem. A first step is to use dags (directed acyclic graphs) sharing common variables or subterms, thus allowing nameless variables. For further developments on term implementation, see [Hul80a].

1.3 EQUATIONAL VARIETIES

The main result in this section is Birkhoff's theorem. It asserts that the problem of the validity of an equation in a variety, a semantic problem, can be resolved by the study of an equational congruence in a free algebra, a syntactic problem. From now on, we write $\mathbf{T}$ for $\mathbf{T}(F,V)$ and abbreviate F-algebras to algebras.

An equation or identity is a pair of terms, noted $M=N$. An algebra **A** validates the equation $M=N$ iff for all morphisms $\varphi : T \to \mathbf{A}$, we have $\varphi(M) = \varphi(N)$ in the carrier A of **A**. Informally speaking, all interpretations of terms M and N in **A** yield the same element of A; in notation we put $\mathbf{A} \models M=N$. Such an interpretation is given by a map from $V(M) \cup V(N)$ to A by Def.1.2.

Definition 1.5

Let E be a set of equations, the variety $\mathbf{M}(E)$ defined by E is the class of algebras that validate all equations in E.

This variety is noted $V_F(E)$ by Cohn [Coh65]. We omit the subscript F, and allow extensions of the operator domain with a set of constants. The equations from E are also called the axioms or laws of the variety. By Th.1.3, the $\mathbf{M}(E)$-free algebra on V exists. It is called the E-free algebra on V. The variety $\mathbf{M}(E)$ validates the equation $M=N$ iff every algebra in $\mathbf{M}(E)$ validates it. In this case, let us write $\mathbf{M}(E) \models M=N$. A set E of equations also defines a congruence on **T**:

Definition 1.6

If E is a set of equations, then the E-equality is the smallest congruence on **T** *generated by the pairs $(\sigma(M),\sigma(N))$ for $M=N$ in E and σ an arbitrary substitution.*

This congruence is noted $=_E$, and is also called the equational theory generated by E. The completeness theorem of Birkhoff [Bir35] states that the relations $\mathbf{M}(E) \models$ and $=_E$ are essentially the same.

Theorem 1.7 (Birkhoff)

Let E be a set of equations, then for all terms M and N, we have $\mathbf{M}(E) \models M = N$ iff $M =_E N$.

Hence, this result reduces the validity problem to a syntactic one, the study of a congruence in **T**. The present survey is devoted to an analysis of a tool for solving this equality problem for finite sets of equations: the computation of canonical representatives in **T** for $=_E$, via reduction rules generated by the defining equations. More precisely, our aim is the study of the **word problem** for an algebra in $\mathbf{M}(E)$ by rewriting systems.

But, before formally presenting this problem, let us define the equational varieties that we shall examine throughout this monograph. In the rest of this paragraph, x,y and z denote three variables. The simplest equational class is the variety of *Semigroups*. They are defined by an associative binary operator, $F = \{*\}$, $E = \{**xyz=*x*yz\}$. In infix and parenthesis notation, the associativity states that

$$(x * y) * z = x * (y * z).$$

The next variety is the class of *Monoids*. A monoid is a semigroup with a constant I, satisfying the identity laws:

$$x * I = x, I * x = x.$$

The constant I is called the neutral or unit element for the operator $*$. The binary operators may also be commutative (or abelian) ones. In which case we say the algebra is commutative. Hence, commutative semigroups (resp. monoids) satisfy the law of commutativity

$$x * y = y * x.$$

The two cases, commutative or not, have in most cases a strongly different behaviour with respect to word problems. The next class of algebras is the class of *Groups*, intensively studied in this monograph. Non-abelian groups are monoids together with a unary operator $^{-1}$, satisfying the following identities:

$$x * x^{-1} = I, \ x^{-1} * x = I.$$

The usual notation for abelian groups as well as for commutative algebras is the additive one: $x+y=y+x$. In this case the constant will be noted 0.

The remaining classes of algebras we shall consider possess two binary operators: one of them defining an abelian group. The first structure is *Rings*: $F = \{ + , * , 0 \}$, where 0 is the constant of the abelian group structure defined by the operator $+$. The two fundamental laws defined upon $*$ are the associativity and the distributivity:

$$x * (y * z) = (x * y) * z,$$

$$x * (y + z) = (x * y) + (x * z), (x + y) * z = (x * z) + (y * z).$$

Of course, the second operator may be commutative, or possess a unit element. Now, we present *Modules*. Strictly speaking, modules are multi-sorted algebras, defined on *two* carriers [Hue80b], with so-called scalars from a ring. However, we want to compute representatives for congruence classes in free algebras on sets of constants. Also the free scalar ring may be embedded in the module (in a precise meaning defined later on). We can restrict ourselves to one-sorted algebras with partial operators. Another possible definition is to interpret scalars as endomorphisms of the module (see [Coh73,p.52]). Thus, a module M is a commutative group whose operator is $+$, and with three partial multiplicative operators $*$, $\times$ and $\oplus$ such that:

$$x \times (y \times z) = (x * y) \times z,$$

$$x \times (y + z) = (x \times y) + (x \times z), (x + y) \times z = (x \times z) \oplus (y \times z).$$

$$0\times x = 0,\ x\times 0 = 0.$$

Finally, if there exists a last operator such that the module is also a ring, then the structure is called an *Algebra* (which is confusing). From this basic definition of modules or algebras, several variations are possible when the scalar ring is unitary or commutative. We now turn to a precise statement of the word problem.

The set $G = \{ M \mid M \in \mathbf{T},\ V(M)=\phi \}$ of closed terms is naturally structured as an algebra. The E-equality in $\mathbf{T}$ restricted to G is a congruence still noted $=_E$. The E-initial algebra $\mathbf{I}(E)$ is $G/=_E$. The validity problem for $\mathbf{I}(E)$ is the validity problem in $\mathbf{M}(E)$ restricted to closed substitutions (those such that, fo all x in V, $V(\sigma(M))=\phi$):

$$\mathrm{I}(E) \models M=N \Leftrightarrow \forall\sigma, \text{ a closed subs., } \sigma(M) =_E N. \tag{V}$$

However, we do not have the equivalent of Birkhoff theorem in this case. For example, let $F=\{a, b, \times\}$, a, b constants, $\times$ of arity 2. With the set E of equations $a\times b = b\times a$ and $x\times(y\times z)=(x\times y)\times z$ where x, y, z are variables, the algebra $\mathbf{I}(E)$ validates the equation $x\times a = a\times x$, but equational reasoning fails to prove this equation; we need induction. However, we are interested, in the initial algebra $\mathbf{I}(E)$, only in equations between closed terms. This problem remains tractable by equational methods and is called the **word problem**. The equivalence (V) becomes

$$\mathrm{I}(E) \models M = N \Leftrightarrow M =_E N \tag{W}$$

More formally, let us define a **presentation** of an algebra by splitting the equations of E into two parts, a set R of closed equations called the **defining relations** and a set of equations still denoted by E.

Definition 1.8

Let F be an operator domain, E a set of equations on $\mathbf{T}(F,V)$. A presentation of an algebra in the variety $\mathbf{M}(E)$ is a pair (C,R) where C is a set of constants called generators such that $C \cap F=\phi$ and R is a set of equations on the free $\mathbf{M}(E)$-algebra on C.

The algebra $\mathbf{A}$ presented by (C,R) is the quotient algebra of $G(F\cup C)/=_E$ by the congruence generated by R.

The word problem for $\mathbf{A}$ is : given two terms M and N in $G(F\cup C)$, decide whether or not the equation $M=N$ is valid in $\mathbf{A}$.

The uniform word problem for $\mathbf{M}(E)$ is : given a presentation (C,R) and two terms M and N in $G(F\cup C)$, decide whether or not the equation $M=N$ is valid in the algebra presented by (C,R).

If C is finite then the algebra **A** is said to be ***finitely generated***, and ***finitely presented*** when both C and R are finite. As shown by the second definition, the study of **A** needs a decision procedure for the congruence $=_E$. This is the subject of the second chapter, where a completion algorithm is defined that tries to exhibit a canonical element in every congruence class of $=_E$. To conclude this chapter, let us recall the word problem in a more fluent phrase:

- Given a presentation of an algebra **A**, find an algorithm deciding the equality of two closed terms representing abstract elements of **A**.

The uniform word problem is quantified over all presentations in a variety, and so is closely related to this variety rather than to a precise presentation:

- Given an equational variety, find an algorithm that, when given a finite presentation of an algebra **A** and a pair of closed terms, decides their equality in **A**.

For both these questions, the problem is said to be solvable if there exists an algorithm; unsolvable otherwise (see [Dav58] for a development of decidability).

Observe that an equational theory defined by a recursive set of equations is always semi-decidable. One can enumerate the list of consequences of the defining equations. Given an equality, we can halt this process as soon as it appears in the list. If the equation is not valid, this process will not halt.

If, on the other hand, we are given a semi-decision procedure enumerating inequalities of terms, running these two procedures simultaneously gives a decision algorithm. This is the case of so-called ***residually finite*** algebras: an algebra **A** of a variety **V** is residually finite iff for all terms M,N such that $M \neq_{\mathbf{A}} N$, there exists a morphism φ from **A** onto a finite algebra **B** of the variety **V** separating these terms, $\varphi(M) \neq_{\mathbf{B}} \varphi(N)$. Also of importance is the case of ***simple algebras***: an algebra **A** is simple iff its only congruences are the trivial one and the total one (resp. ϕ and $A \times A$ when considered as subsets of $A \times A$, A the carrier of **A**). The only quotient algebras of **A** are itself and the singleton algebra. For both notions, there exist semi-algorithms enumerating the set of equalities and the set of inequalities for a given finitely generated recursively related algebra. Let us describe them roughly (see [Eva78a], we assume the algebras recursively presented). For inequalities in the residually finite case, just enumerate the finite algebras searching one that separates the inequality's members. For simple algebras, if $M \neq_{\mathbf{A}} N$, then the congruence generated by the pair (M,N) is $A \times A$. If we know two terms M_0,N_0 with $M_0 \neq_{\mathbf{A}} N_0$, then enumerating the equalities obtained from **A**'s presentation increased by $M=N$ will eventually produce $M_0=N_0$, as **A** is simple. For a converse stating that an algebra has a solvable word problem iff it is embeddable in a finitely generated simple algebra see [Eva78b]. Thus running these semi-algorithms simultaneously yields a decision procedure for the word problem of the algebra.

To conclude this chapter, we define briefly regular and context-free languages and the regular expressions, used essentially in chapter 6 for the description of the set of irreducible terms. For a complete development, see [Hop79]. The set Σ^* of *strings* over an *alphabet* Σ is the free monoid on Σ. A *language* is a subset of this monoid. Given two languages L_1 and L_2, their concatenation $L_1.L_2$ is the set $\{ xy \mid x \in L_1, y \in L_2 \}$. If the empty string or the neutral element for the concatenation is denoted by ε, we set, for a language L:

i) $L^0 = \{\varepsilon\}$.

ii) $L^i = L.L^{i-1}, i>0$.

iii) $L^* = \bigcup_{i=0}^{\infty} L^i$, the Kleene closure of L.

Now, we define the algebra of regular expressions on the alphabet Σ to be the free F-algebra on Σ, where F contains the following five operators:

i) The constants ϕ and ε,

ii) The unary operator $*$,

iii) The two binary operators . and +.

The intended meaning is that the constants ϕ and ε respectively denote the empty language and the language restricted to the empty string singleton. If expressions l_1 and l_2 denote the languages L_1 and L_2, then the expressions l_1+l_2, $l_1.l_2$ and l_1^* respectively denote the union, the concatenation of L_1 and L_2, and the Kleene closure of L_1. A language is regular iff there exists a regular expression denoting it. In describing normal forms, we use freely the well-known correspondence between regular expressions and languages accepted by a finite automaton [Hop79,pp.30-34].

The context-free languages appear in chapter 3 as congruence classes of reduction relations. A context-free grammar G is composed of three finite sets:

i) Σ, the alphabet of *terminals*,

ii) V the alphabet of variables or non-terminal symbols, with a distinguished element s, the start symbol, and

iii) R the set of *productions* or pairs (x,y) with x in V and y in $(V \cup \Sigma)^*$.

The transitive-reflexive closure $\xrightarrow[R]{*}$ of the set R considered as a relation defines a language $L(G)$ on the alphabet Σ by

$$\forall w \in \Sigma^*, w \in L(G) \text{ iff } s \xrightarrow[R]{*} w.$$

A language is context-free iff it is the language $L(G)$ for some context-free grammar G. The class of context-free languages is exactly the class of languages accepted by pushdown automaton [Hop79,pp.114-119].

2 Canonical forms

This chapter presents a model of computations in term algebras: the rewriting systems. This theory is still an area of intensive research. Recent developments are mentioned throughout the chapters with references. A first section is devoted to unification or the problem of solving equations in a free algebra. Then, it will be possible to define the basic notions of rewriting and present the completion algorithm. A third section details the more tedious case of the commutative-associative theories (this section may be skipped in a first reading). The application of the method to classical algebra is the subject of the last section.

2.1 SOLVING EQUATIONS

In the previous chapter, we mentioned the validity problem for an equation in an algebra. On the other hand, given an equation, we may ask for substitutions that, when applied to this equation, yield a valid one. This is the **unification** problem introduced by J. Herbrand [Her30], recognized as fundamental in theorem proving by J. Robinson [Rob65], and formally stated as follows. Let as previously E and V be a finite set of equations and a denumerable set of variables, the operator domain being implicit. In the E-free algebra on V, given two terms M and N, find when they exist all endomorphisms σ such that $\sigma(M)=_E \sigma(N)$. From now on, the sets F and V are supposed to be constant; we omit their explicit mention.

Definition 2.1

Two terms M and N in $\mathbf{T}$ are E-unifiable when there exists a substitution σ such that $\sigma(M)=_E\sigma(N)$.

Such a substitution is called a E-**unifier** of M and N. The unification problem is the construction of all those unifiers. In general they form an infinite set as when σ is a solution, then for every substitution τ, their composition $\tau\circ\sigma$ is a solution. We are looking for a generating subset of unifiers, or even better, a minimal generating set or basis of E-unifiers. Also we introduce a partial-order (transitive and reflexive relation) on substitutions:

Definition 2.2

Let V *be a finite set of variables,* σ, τ *two substitutions, we define* $\sigma \leq_{E,V} \tau$ *iff there exists a substitution* ρ *such that :*

$$\forall x \in V \quad \tau(x) =_E \rho(\sigma(x))$$

Let $U_E\ (M,N)$ be the set of all E-unifiers of the terms M and N.

Definition 2.3

Let M,N *be two terms,* $V = V(M) \cup V(N)$, *and* W *be a finite set of variables. A set* Σ *of substitutions is a complete set of E-unifiers of* M *and* N *away from* W *iff*

$$i)\ \forall \sigma \in \Sigma,\ D(\sigma) \subset V\ \&\ W \cap (\bigcup_{x \in D(\sigma)} V(\sigma(x))) = \phi,$$

$$ii)\ \Sigma \subset U_E\ (M,N),$$

$$iii)\ \forall \sigma \in U_E\ (M,N), \exists\, \rho \in \Sigma\ s.t.\ \rho \leq_{E,V} \sigma,$$

Moreover, Σ *is minimal if*

$$iv)\ \forall \sigma, \tau \in \Sigma, \sigma \neq \tau \Rightarrow \sigma \not\leq_{E,V} \tau.$$

The unification appears in the work of Herbrand and was rediscovered by Robinson [Her30, Rob65]. This definition is inspired from Plotkin and Huet [Plo72, Hue76]. The set W of variables is needed for technical reasons, namely the problem of α-conversion (cf.§.1.2). We said in the previous chapter that two terms were equivalent when they were equal modulo a permutation of their variables. They are said to be α-convertible. A good unification algorithm must solve this problem in a pleasant way, see [Hul80a, Fag83b]. For our problem, when computing a set Σ, we have to protect the variables already substituted with the set W. Now, we examine some cases according to the set E (for proofs of unification algorithms, we refer to the literature).

First of all, when $E = \phi$. There exists a minimal unifier when the two terms are unifiable, which forms a basis [Rob65]. This case of great practical importance for Prolog-like languages and the resolution principle or for rewriting systems has been intensively studied. For further information, see [Fag83b, Mar82, Pat78, Rob71]. Let us present an algorithm due to Huet [Hue76]. Its structure will be used later on: it loops on two basic operations: the former is a decomposition of terms, the latter builds the unifier.

A **unificand** is a finite set of term pairs. Let $\Phi=\{(M_i,N_i)/i=1...n\}$ be a unificand; its set of variables is $V(\Phi) = \bigcup_{i=1}^{n} V(M_i)\cup V(N_i)$. A substitution σ is a unifier of Φ iff $\sigma(M_i) = \sigma(N_i)$, $1\le i\le n$. The set of all Φ-unifiers is noted $U(\Phi)$. The first step is performed by the following algorithm *Simpl* (Φ) :

Simpl(Φ) = **While** $\exists(M,N)\in\Phi$ *such that* $M\notin V$ *and* $N\notin V$

Do { $M=fM_1 \cdots M_n$ *and* $N=gN_1 \cdots N_m$, $f,g\in F$

If $f\neq g$ **Then** *Fail*

Else $\Phi:=[\Phi-\{(M,N)\}]\cup\{(M_i,N_i) \mid i=1...n\}$ };

Replace all pairs (M,x) *in* Φ *by* (x,M), $M\notin V$, $x\in V$;

Delete all pairs (x,x) *in* Φ, $x\in V$;

If *there exists a pair* (x,M) *in* Φ *with* $x\in V(M)$

Then *Fail*

Else *Return* Φ ∎

An induction on the size of terms proves the termination of *Simpl*. Moreover we have the relations $U(Simpl(\Phi))=U(\Phi)$ and $V(Simpl(\Phi))\subseteq V(\Phi)$.

Unification Algorithm

Input: *Two terms* M *and* N.

$\Phi_1:=Simpl(\{(M,N)\})$; $i:=1$; $\sigma_1:=\phi$;

Loop { **If** $\Phi_i=\phi$ **Then** *return* $\sigma_i\circ\sigma_{i-1}\circ\cdots\circ\sigma_1$;

If $\Phi_i=Fail$ **Then** *Fail*;

Let (x_i,M_i) *in* Φ_i, $x_i\in V$;

Let σ_i *be the substitution* (x_i,M_i);

$\Phi_{i+1}:=Simpl(\{(\sigma_i(M_j),\sigma_i(N_j)) \mid (M_j,N_j)\in\Phi_i-\{(x_i,M_i)\}\})$;

$i:=i+1$ } ∎

The termination follows from the fact that the cardinality of $V(\Phi_i)$ strictly decreases at each iteration. Then, the proof of accuracy consists in proving that the substitution computed is the main unifier of M and N [Hue76]. In practice, this algorithm is expensive both in time and memory, due to the copies of terms done by the simplification. Huet gives an algorithm working on shared data-structures (dags) with the double advantage of being quasi-linear in time and memory and working as well on infinite regular terms. For further developments see [Fag83a, Mar82, Pat78, Rob71]. The previous algorithm is sufficient for a presentation of the theory. A recent result shows that unification, due to the loop on simplification and construction of the substitution, is a sequential algorithm. No significant gain in efficiency results from a parallel unification algorithm [Dwo83].

Together with the unification problem, we need a procedure that matches a term P with another one Q, a match from P to Q being a substitution σ such that $\sigma(P) =_E Q$. When E is the empty set, this algorithm is quite simple:

Matching Algorithm

Input: Two terms P,Q;

$\sigma := \phi$;

Match(P,Q) where

Match(M,N) := If M = f $M_1 \cdots M_n$ then

{If N = f $N_1 \cdots N_n$ then

for i from 1 to n Match(M_i,N_i);

return σ;

else Failure}

else M∈V

{If $\sigma(M)$ is undefined then $\sigma := \sigma.<M,N>$

else if $\sigma(M) \neq N$ then Failure ▪

This algorithm computes the partial order ≤ on terms defined in the previous chapter. It is linear in the size of Q, and does not need multiple term copies. Moreover, its structure is simpler than the unification algorithm: we have no loop on the simplification and construction steps.

When E is reduced to the associativity law of a binary operator, a complete set of unifiers always exists, but it may be infinite: if $M=a.x$ and $N=x.a$, $x \in V$, a a constant, then the set $\{<x,a.(a\ldots(a.a)\ldots)>\} \cup \{<x,a>\}$ is an infinite set of incomparable unifiers. This is the hard problem of equations with constants in the free monoid, recently proved decidable by Makanin [Mak77b] with a complex algorithm [Péc81]. Makanin has also proved that group unification is decidable [Mak82], with an algorithm even more complex than that of monoid unification.

The commutative case has been studied by Siekmann [Sie79] and Lankford [Lan77a]. There always exists a finite minimal set of unifiers [Sie84]. Livesey, Siekmann, Szabo and Unvericht proved that unification in presence of associative-distributive operators is undecidable [Liv79]. Lankford has also studied the abelian group unification [Lan79], Livesey and Siekmann the combination of associativity, commutativity and idempotence [Liv76]. For the problem of unification in higher order languages, see [Gol81, Hue76, Jen77]. The problem is undecidable at second order.

If E consists in the associativity and commutativity laws of one or several binary operators, then there exists a finite minimal set of unifiers. Stickel [Sti81] proposed an algorithm computing such a set, including the case where F contains other operators in addition to the associative-commutative ones. We study this case in the third section, after the presentation of the rewriting theory.

2.2 REWRITING SYSTEMS AND THE COMPLETION ALGORITHM

In a first section we shall introduce rewriting rules, with some of their properties and theorems; mainly the **confluence** (or Church-Rosser property) with the Newman lemma and the Knuth-Bendix theorem. A second section is devoted to the **Completion Algorithm** and contains a detailed proof of its correctness. This proof is a fundamental result of which the corollaries will be developed in the next two chapters.

2.2.1 Rewriting systems

The goal of the method is to provide a decision procedure for the validity problem in an equational variety defined by a finite set E of equations, by the study of the E-equality in the free algebra $\mathbf{T}(F,V)$, for these two problems are equivalent from the Birkhoff Completeness theorem (Thm.1.7). This goal is achieved as soon as we can exhibit a canonical representative for each congruence class of the equality such that every element in the class ***reduces*** on this representative. Thus, an equation $M=N$ holds in the variety iff the canonical representatives of M and N, called their ***canonical form***, are syntactically equal, the set of canonical forms is a transversal in the sense of Cohn [Coh65]. The reduction is performed with ***rewriting rules*** computed from equations in E.

Definition 2.4

A rewriting rule is a pair of terms (M,N) such that $V(N) \subseteq V(M)$. The term P rewrites in the term Q iff there exists an occurrence u in $O(P)$ and a substitution σ such that $\sigma(M) = P/u$ and $Q = P[u \leftarrow \sigma(N)]$.

A set of rules will be represented by R or S. In notation, we put $P \xrightarrow[R]{} Q$. The set R defines by matching a relation on $\mathbf{T}(F,V)$. The relation $\rightarrow$ (we omit the subscript R when no confusion is possible) satisfies the following properties:

1) Stability: $\forall M,N\ \forall\sigma,\ M \rightarrow N \Rightarrow \sigma(M) \rightarrow \sigma(N)$,

2) Compatibility: $\forall P,\ \forall u \in O(P),\ M \rightarrow N \Rightarrow P[u \leftarrow M] \rightarrow P[u \leftarrow N]$.

In fact, the relation $\rightarrow$ is the smallest one on $\mathbf{T}$ satisfying these two properties. Moreover, in order to use this relation mechanically, i.e. apply the rules of reductions until an irreducible term is reached, then we need the two following properties:

3) Termination: there exists no infinite reduction chain starting at a term M. Or equivalently, the relation $\rightarrow$ is well-founded.

4) Confluence: $\forall M, M_1, M_2,\ M \xrightarrow{*} M_1\ \&\ M \xrightarrow{*} M_2 \Rightarrow \exists P\ M_1 \xrightarrow{*} P\ \&\ M_2 \xrightarrow{*} P$.

The noetherianity implies the existence for all terms of at least one irreducible or normal form. Of course, this normal form is not necessarily unique. It is the case when the rewriting system is confluent. The unique normal form of M is then called the canonical form of M, and is noted $R(M)$. This method of decision for an equality is of great importance, not only for the first order terms, but also for the λ-calculus [Chu36, Cur58, Bar80] or in proof theory [Bar77].

Thus, we want those sets of rules having properties 3) and 4). The noetherianity is checked by partial-orderings (transitive and irreflexive relations) on terms. We shall see several such orderings, they have to satisfy at least the following three properties:

Definition 2.5

A partial-ordering $>$ on terms is a reduction ordering iff, for all terms M,N and P, we have

i) *Stability,* $M > N \Rightarrow \forall\sigma,\ \sigma(M) > \sigma(N)$.

ii) *Compatibility,* $M > N \Rightarrow P[u \leftarrow M] > P[u \leftarrow N],\ \forall u \in O(P)$.

iii) *Noetherianity, there is no infinite decreasing sequence.*

So, given a set of rules R and a reduction ordering $>$, if $M \rightarrow N \in R$ implies $M > N$, then R is trivially noetherian. Conversely, if R is noetherian, then the relation $\xrightarrow[R]{+}$ is a reduction ordering. Although there is an intensive research on this subject of rewriting termination, we need some classical orderings, presented in the last section of this chapter. We do not require a total ordering; a partial one is sometimes sufficient.

For the presentation of the Completion procedure or Knuth-Bendix algorithm, we assume that all reductions are terminating ones. However, the confluence is a global property and is thus hard to check mechanically. The following property is the localization of the confluence:

Definition 2.6

A rewriting system is locally confluent iff

$$\forall M, M_1, M_2\ M \longrightarrow M_1\ \&\ M \longrightarrow M_2 \Rightarrow \exists P\ M_1 \xrightarrow{*} P\ \&\ M_2 \xrightarrow{*} P$$

Then, we have the classical lemma due to Newman [New42]:

Lemma 2.7 (Newman)

Let R be a noetherian rewriting system, then:

R is confluent iff R is locally confluent.

Of course, this lemma is true for every noetherian relation on an arbitrary set. A simple proof using the principle of noetherian induction can be found in [Hue80a]. This induction is sketched on the following figure. If $M=M_i$, $i=1$ or 2, then the result is trivial, otherwise there exist A and B corresponding to the following diagram:

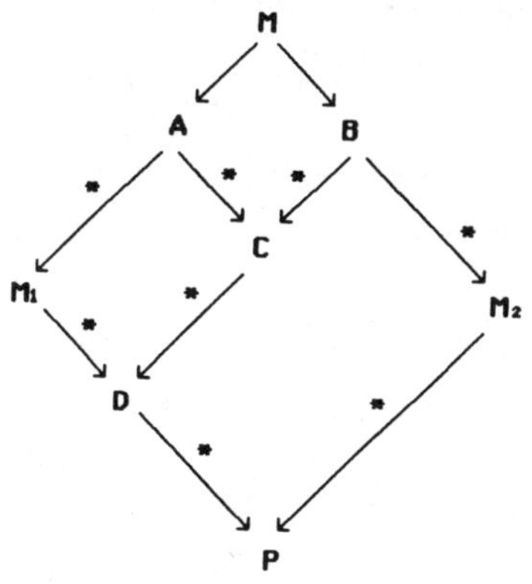

Fig. 2.1

Under a confluent and noetherian set of rules R, every term M possesses a unique normal form, noted $R(M)$. The local confluence is still too global to be computerized: it is quantified over all terms, but now we may use the fact that our relation is finitely generated by some rules. With the property of termination, the non-confluence must reduce to some elementary divergence between rules. Indeed, this is true. These divergent points are called critical pairs.

Definition 2.8

Let $M \rightarrow N$ and $P \rightarrow Q$ be two rules in R, and u be an occurrence in $O(M)$ such that $M/u \notin V$, M/u and P are unifiable, let σ be a main unifier of M/u and P such that $V(\sigma(M)) \cap V(M) = \phi$. If μ is the match from M/u to $\sigma(P) = \sigma(M/u)$ and ν the match from P to $\sigma(P)$, then the critical pair obtained by the superposition of the rule $P \rightarrow Q$ on the rule $M \rightarrow N$ at occurrence u is the pair

$$(\mu(M)[u \leftarrow \nu(Q)] \, , \mu(N)).$$

The condition on variables prohibits a possible link created by the unification between variables of M and P: the sets of left member variables are disjointed, as the rules express universally quantified identities of the E-equality. This bind would not give the most general unifier: if $x.y \rightarrow y$ and $z.f(x) \rightarrow z.x$, then the most general critical pair is computed by two reductions from $x.f(y)$, not from $x.f(x)$. The local confluence of R needs the confluence of the critical pair members. In fact, the converse is true as first shown by Knuth and Bendix [Knu70]:

Theorem 2.9 (Knuth-Bendix)

A rewriting system R is locally confluent iff

$$\forall (P,Q) \text{ critical pair of } R, \exists N \; P \overset{*}{\underset{R}{\Rightarrow}} N \;\&\; Q \overset{*}{\underset{R}{\Rightarrow}} N.$$

In [Knu70], the proof uses the assumption of noetherianity. In fact this hypothesis is redundant as shown in [Hue80a], the implication

critical pairs are confluent $\Rightarrow$ loc. confluence

is the non-trivial part and is proved by case analysis. The proof is technically tedious but conceptually simple. We sketch it in three diagrams on next page.

Thus, only the last case involves critical pairs. In other cases, the confluence of M_1 and M_2 is straightforward, but some care is needed in a formal proof concerning the occurrences and the different substitutions. We point out that these three cases will occur in other places in this work, especially with the next theorem, and in chapter 6. Let M, M_1, M_2 be such that $M \underset{R}{\Rightarrow} M_1$ and $M \underset{R}{\Rightarrow} M_2$, we have three cases:

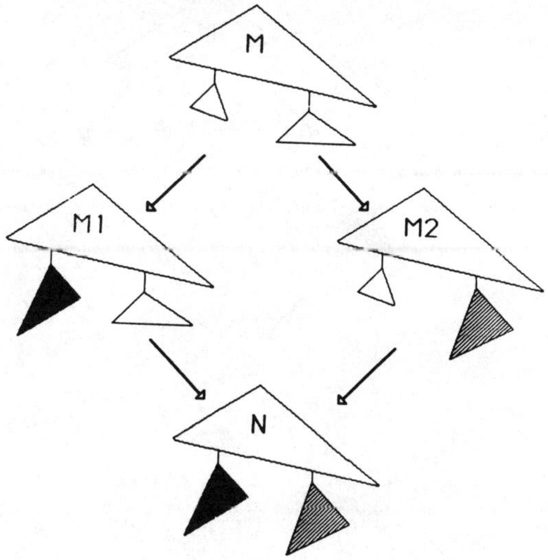

Fig. 2.2: Disjoint Reductions

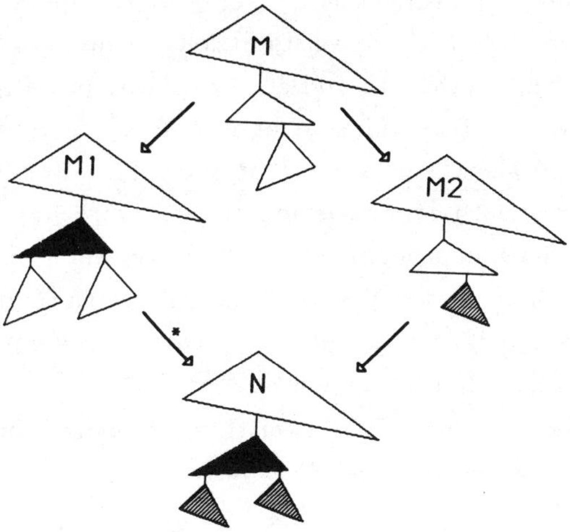

Fig. 2.3: Nested Reductions

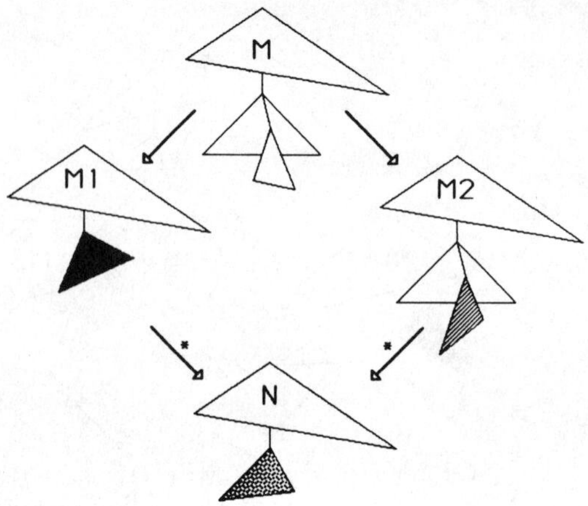

Fig. 2.4 : Overlapping Reductions

2.2.2 The completion algorithm

The Knuth-Bendix theorem gives an algorithm testing the local confluence of a set of rules via the critical pairs. Such a pair is said to be confluent when its two members reduce to the same term under a rewriting system. Also when a set of rules is not locally confluent, there exists at least a non-confluent critical pair between two rules. The hint is then to add a new rule by orienting this pair. This cancels a case of divergence. Thus, when a non-confluent pair is found, it is added to the current set of rules. Given a set E of equations over $\mathbf{T}(F,V)$, the completion procedure orientates both equations and critical pairs, according to a reduction ordering. This compilation process may halt as an equation cannot be oriented, loop undefinitely, or stop in success when all critical pairs are resolved (are confluent). In this last case, the set of rules is locally confluent and noetherian and thus confluent. We call such a set **complete**.

In the presentation of the algorithm, rules are labelled by integers, and noted $k:\lambda \rightarrow \rho$. They are marked when they have been selected for superposition.

Completion Algorithm

Input E : A finite set of equations,

$<$: A reduction ordering,

$Super(k,R)$: Computes critical pairs between rule k and

rules in R with label less or equal than k.

$R_0 := \phi$; The set of rules.

$E_0 := E$; The set of waiting equations.

$i := p := 0$; The step counter and the rules counter.

Loop { **If** $E_i \neq \phi$ **Then** { *choose a pair* $(M_0, N_0) \in E_i$; *Create_Rule*$(R_i(M_0), R_i(N_0))$ }

Else { **If** *all rules are marked* **Then** *Success*

Else { *choose an unmarked rule* k;

$E_{i+1} := Super(k, R_i)$; $R_{i+1} := R_i$;

mark rule k; $i := i+1$ }}}

where Create_Rule(M, N) =

If $M = N$

Then { $E_{i+1} := E_i - \{(M_0, N_0)\}$; $R_{i+1} := R_i$; $i := i+1$ }

Else { **If** $M > N$ **Then** { $\alpha := M$, $\beta := N$ }

If $M < N$ **Then** { $\alpha := N$, $\beta := M$ }

Else *Fail*;

Let K *be the set of rules in* R_i *whose left member is reducible by the rule* $\alpha \rightarrow \beta$;

$E_{i+1} := E_i \cup \{ (\lambda, \rho) \mid k : \lambda \rightarrow \rho \in K \}$;

$R_{i+1} := \{ k : \lambda \rightarrow \rho' \mid k : \lambda \rightarrow \rho \in R_i - K,\ \rho' = (R_i \cup \{\alpha \rightarrow \beta\})\ (\rho) \}$

$\cup \{ p+1 : \alpha \rightarrow \beta \}$;

$p := p+1$; $i := i+1$ ∎

From a practical point of view, it is useful to keep the set K whose members are the reduced rules, and to select pairs in K before those in E_i, they frequently give simpler rules. Two instructions include a choice of a rule or an equation. For this purpose, a size function on terms may be used. This fact has two consequences: first, the efficiency is increased; second, such a function allows the following **fair selection hypothesis**:

$\forall i$, $\forall k : \alpha \rightarrow \beta \in R_i$, $\exists j \geq i$ such that
either the rule k is suppressed at the jth iteration,
or the rule k is selected for superposition at the jth iteration.

We may then assert the correctness of this algorithm [Hue81]. This theorem will be used many times in the present study: it may be seen as a result of universal algebra whose instanciations are analyzed in this survey. Consequently, we present a proof of this result.

Theorem 2.10

Suppose that the fair selection hypothesis holds and that the completion procedure does not loop undefinitely.
Let $R_\infty = \{ k:\alpha \longrightarrow \beta \mid \exists i,\ k:\alpha \longrightarrow \beta \in R_i \ \&\ \forall j \geq i,\ k:\alpha \longrightarrow \beta \in R_j \}$ *then* $M =_E N \Leftrightarrow M =_{R_\infty} N$. *Moreover,*

—If $k:\lambda \longrightarrow \rho \in R_\infty$, *then* λ *and* ρ *are in* $R_\infty - \{ k:\lambda \longrightarrow \rho \}$*-canonical form.*

—If R_∞ *is finite, then the algorithm halts after a finite number of iterations.*

The proof is essentially Huet's one. It proceeds in three steps. Let $R = \bigcup_{i=1}^{i=\infty} R_i$.

Lemma 2.11

The equality $=_R$ *generated by the rewriting system* R *is the same as* $=_E$.

Proof. We may first observe that

$$\forall\ \lambda \longrightarrow \rho \in R_i\ ,\ \forall\ j > i\ , \exists\ \alpha \longrightarrow \beta \in R_j \text{ such that } \alpha \longrightarrow \beta \text{ reduces } \lambda.$$

As when a rule is deleted, the new rule introduced at this step reduces its left member. We conclude that if the term M is R_i-reducible, then it is R_j-reducible. Note, however, that we do not have $\underset{R_i}{\longrightarrow} \subset \underset{R_j}{\longrightarrow}$.

Now, we have $=_{\mathbf{R}} \subset =_E$, as all critical pairs thus all rules are consequences of equations in E. In order to prove the remaining inclusion, it is sufficient to show that

$$\forall\ i \geq 0,\ \forall\ (M,N) \in E_i\ \exists P \text{ such that } M \overset{*}{\underset{R}{\longrightarrow}} P\ \&\ N \overset{*}{\underset{R}{\longrightarrow}} P. \tag{1}$$

Observe that $=_E \subset =_{\mathbf{R}}$ follows by putting $i=0$. If the pair (M,N) is selected for a possible creation of a rule, then the previous assertion is obvious. Thus, we must prove that every such pair is selected at some stage. This in turn is achieved if :

$$\forall\ i \geq 0,\ E_i \neq \phi \Rightarrow \exists\ j > i \text{ such that } E_j = \phi.$$

Let us now consider the integer $n = |E_i| + |R_i|$. At each call of *Create_Rule*, either it decreases (no rule is created) or remains constant (a rule is created). In this last case we look at the $2n$-tuple of left and right members ordered componentwise by the reduction ordering. As for the integer n, this vector strictly decreases when a reduction is performed. Otherwise it remains constant, but the integer $|E_i|$ decreases itself, as K is empty. Thus the set E_i is necessarily empty at some iteration following the ith-one. This concludes the proof of lemma 2.11.

Now, let us point out that *Create_Rule* keeps the rules of R_i inter-reduced. We say that R_i is in reduced form. It follows from the construction of R_∞ that this set is in reduced form, as two distinct rules belong to some R_k for k sufficiently large.

Now, our aim is to prove $=_{R_\infty} \equiv =_E$. As the inclusion $=_{R_\infty} \subset =_R$ is trivial, it suffices to prove the other inequality. In so doing, we shall prove that R_∞ is noetherian and confluent. But we need a hypothesis under which no rule will be forgotten for the computation of critical pairs:

Hypothesis: For every rule label k, there exists an index of iteration i such that either rule k is selected for superposition or is deleted from R_i.

This is easily achieved by a complexity measure on the rule left members for example. We first prove that given two R_∞-reductions of a term M. The relation R may be used to close the diagram.

Lemma 2.12

For all terms M,N_1,N_2 *such that* $M \xrightarrow[R_\infty]{} N_1$ *and* $M \xrightarrow[R_\infty]{} N_2$,

there exists three terms P_1, P_2, Q *with*

$M \xrightarrow[R]{} P_1 \xrightarrow[R]{*} N_1$, $M \xrightarrow[R]{} P_2 \xrightarrow[R]{*} N_2$ *and* $P_i \xrightarrow[R]{*} Q$, $i=1,2$.

This lemma is sketched by the following diagram:

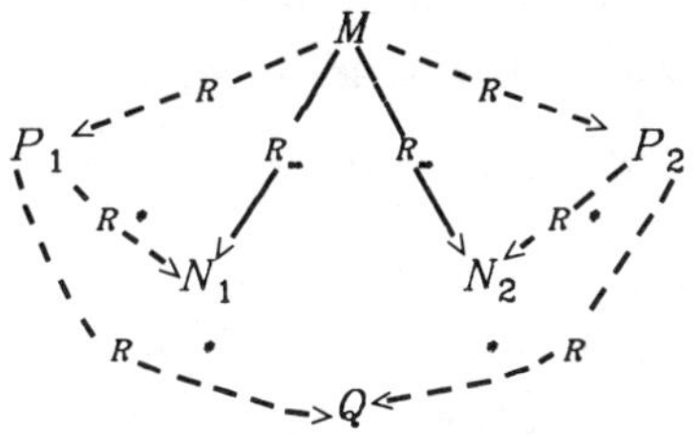

Fig. 2.5

Proof. The proof is by inspection on occurrences of the redexes in M. In the first two cases of the Knuth-Bendix theorem, when the redexes share no symbol, we put $P_i=N_i$, $i=1,2$, and we get easily the term Q. Otherwise, we have two rules $k:\lambda \rightarrow \mu$ and $l:\rho \rightarrow \nu$ in R_∞, with $k \leq l$, $N_1=M[u_1 \leftarrow \sigma_1(\mu)]$ and $N_2=M[u_2 \leftarrow \sigma_2(\nu)]$. As these rules belong to R_∞ and share common symbols, there exists by the Hypothesis an iteration n such that the rule l is selected for the superpositions. Note that at this iteration

n, the right members of the rules are μ' and ν', possibly distinct from μ and ν. However, there exist a permutation σ and a critical pair (C_1,C_2) in E_{n+1} such that, if $u=\min(u_i)$, then we have $P_1 =_{def} M[u_1 \leftarrow \sigma_1(\mu)]$, $P_2 =_{def} M[u_2 \leftarrow \sigma_2(\nu)]$, and $P_i/u=\sigma(C_i)$. But the proposition (1) states the existence of a term O with $C_i \xrightarrow[R]{*} O$, then $P_i \xrightarrow[R]{*} Q$ if $Q=P_i[u \leftarrow O]$. Also, as $\mu' \xrightarrow[R]{*} \mu$ and $\nu' \xrightarrow[R]{*} \nu$ we get $P_i \xrightarrow[R]{*} N_i$ ∎ We now show that the two congruences $=_{R_\infty}$ and $=_R$ are equal. At the same time, we establish the confluence of these two rewriting systems.

Proposition 2.13

For all terms M, N, N_1, N_2, we have

a) $M \xrightarrow[R]{+} N \Rightarrow \exists P\,(M \xrightarrow[R_\infty]{+} P \;\&\; N \xrightarrow[R_\infty]{*} P)$

b) $M \xrightarrow[R_\infty]{*} N_i \Rightarrow \exists P\; N_i \xrightarrow[R_\infty]{*} P$, i=1,2.

c) $M \xrightarrow[R]{*} N_i \Rightarrow \exists P\; N_i \xrightarrow[R_\infty]{*} P$, i=1,2.

Proof. The three implications state that R_∞ closes three diagrams, the second one being the confluence of R_∞. They will be proved simultaneously by structural induction on the reduction ordering: assume a), b) and c) for all terms M' with $M>M'$.

Let $M \xrightarrow[R]{} M_1 \xrightarrow[R]{*} N$ be the reductions of M in a), the first one being done by the rule $k:\lambda \rightarrow \mu \in R$. We prove a) by induction on λ using the well-founded ordering $\gg$ composed of the strict parts of both the subterm and the instanciation orders (cf.§1.2).

If there exists a term μ' such that $k:\lambda \rightarrow \mu' \in R_\infty$, then with rule k, we have $M \xrightarrow[R_\infty]{} M_2$. We have also $\mu \xrightarrow[R]{*} \mu'$, which implies $M_1 \xrightarrow[R]{*} M_2$. Let us apply the induction hypothesis c) to $M_1 \xrightarrow[R]{*} N, M_2$. We find a term P such that $M \xrightarrow[R_\infty]{+} P$ and $N \xrightarrow[R_\infty]{*} P$.

Otherwise, at some iteration n, the rule $k:\lambda \rightarrow \mu'$ is reduced in its left hand side by a new rule $l:\rho \rightarrow \nu$, which reduces M into M_2, while as previously $\mu \xrightarrow[R]{*} \mu'$ yields a term M_3 with $M_1 \xrightarrow[R]{*} M_3$. But the terms M_i, i=2,3, differ from one another in the equation from E_{n+1} $\lambda[u \leftarrow \sigma(\nu)]=\mu'$, so that proposition (1) gives a term N' with $M_i \xrightarrow[R]{*} N'$, i=2,3. Now we apply induction hypothesis c) to the triple M_1,N and N'. This gives O with $N \xrightarrow[R_\infty]{*} O$ and $N' \xrightarrow[R_\infty]{*} O$. Eventually, as $\lambda \gg \rho$, we use induction hypothesis a) on the reduction $M \xrightarrow[R]{+} O$ (remember that $R_\infty \subset R$). This yields the desired term P, with $M \xrightarrow[R_\infty]{*} P$ and $N \xrightarrow[R_\infty]{*} P$.

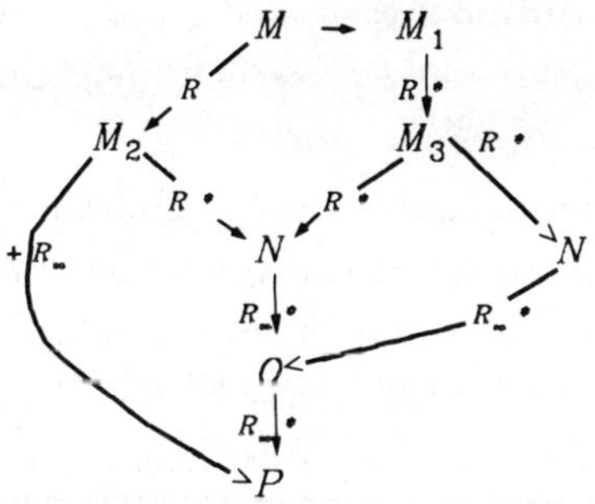

Fig. 2.6

We use lemma 2.12 to show b). First, when M equals N_1 or N_2, we have $M \xrightarrow[R_\infty]{} M_i$, $i=1,2$. Second, this lemma gives three words P_j, $j=1,2$, and Q, such that $P_i \xrightarrow[R]{*} Q$, $P_i \xrightarrow[R]{*} M_i$, $i=1,2$. We apply induction hypothesis c) first to P_1, M_1 and Q to get a term P', second to the triple P_2, M_2 and P' to conclude the existence of the desired term P, as shown in figure 2.7:

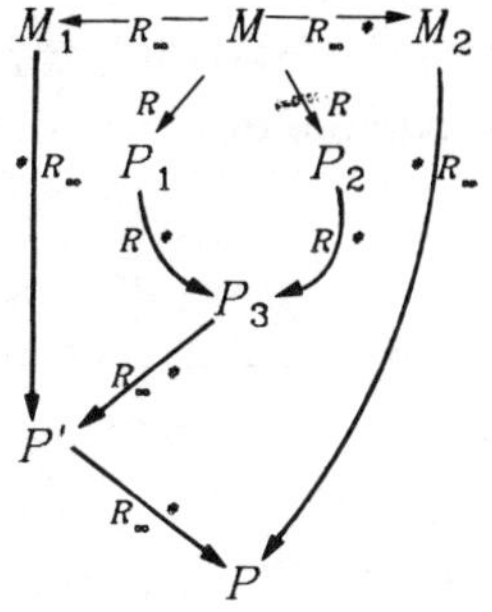

Fig. 2.7

Then we conclude by the proof of c), straightforward with cases a) & b). Once more, if M equals one of the terms N_i, $i=1,2$, then the result is obvious. Otherwise we are allowed to use twice the assertion a) on terms M, M_i. So that $M, N_i \xrightarrow[R_\infty]{*} P_i$, $i=1,2$. Afterwards b) applied to M, P_1, P_2 yields the term P with $N_1 \xrightarrow[R_\infty]{*} P$ and $N_2 \xrightarrow[R_\infty]{*} P$ ∎

With these lemmas, the proof of theorem 2.10 is at hand. The set R_∞ is noetherian, inter-reduced and confluent. From the last assertion of Prop.2.13, we obtain that $=_{R_\infty} \equiv =_R \equiv =_E$. Thus the completion is a semi-decision procedure: if $M =_E N$, then there exists an iteration n such that the R_n-normal forms of M and N are syntactically equal. Now, if R_∞ is finite, then there exists an iteration n such that

$R_n = R_\infty$, as no other rules can be introduced and R_∞ provides a decision procedure for $=_E$. All the waiting equations in E_n are confluent and the algorithm halts after checking this confluence ▪

In other words, with an adequate ordering and if selections are correctly performed, then we get a complete set of rules, possibly infinite. Of course, there exist infinite R_∞ recursively enumerable non-recursive, which surely occur with undecidable theories, while others, still infinite, may be finitely coded. Let us recall that the rules in R_∞ are those computed at an iteration i and whose both left and right members are never reduced by the following iterations, and that the rules are constantly inter-reduced. This is a major point in a completion. Of course for efficiency, but also for the proof of Thm.2.10 as well as for future proofs and for conciseness and elegance of the complete sets. Moreover, Métivier [Mét83] has shown that for a given reduction ordering, if such a reduced complete system exists it is unique. Buchberger and Winkler [Win83] proposed an improvement of the algorithm in the critical pair computation. Informally, if the critical term $\sigma(P)$ of Def.2.8 has a path from its top to the superposition occurrence made of several critical terms, then the computation of the associated critical pair is obsolete. It is not obvious, however, if such a theoretical improvement is a practical one. This improvement of the Knuth-Bendix theorem is in fact a consequence of the generalized Newman lemma [Win83]:

Lemma 2.14

Let $>$ and $\rightarrow$ be two relations on a set S such that $>$ is noetherian and $\rightarrow \subset >$, then the relation $\rightarrow$ is confluent iff

$\forall M, M_1, M_2 \in S,\ M \rightarrow M_1 \ \&\ M \rightarrow M_2$

$\Rightarrow \exists N_i,\ i=1,\ldots,n$ *s.t* $M > N_i,\ M_1 = N_1,\ M_2 = N_n$, *and*

$N_i \rightarrow N_{i+1}$ *or* $N_{i+1} \rightarrow N_i,\ i=1,\ldots,n-1.$

A proof of this lemma can be found in [Win83, Win84]. It corresponds to Thm4.9,p.25 in Cohn [Coh65], where it is called the Diamond lemma, and is implicit in Bergman's paper [Ber78]. The key is that the whole chain of the N_is stays below M. We have the following roller-coaster diagram allowing easy induction:

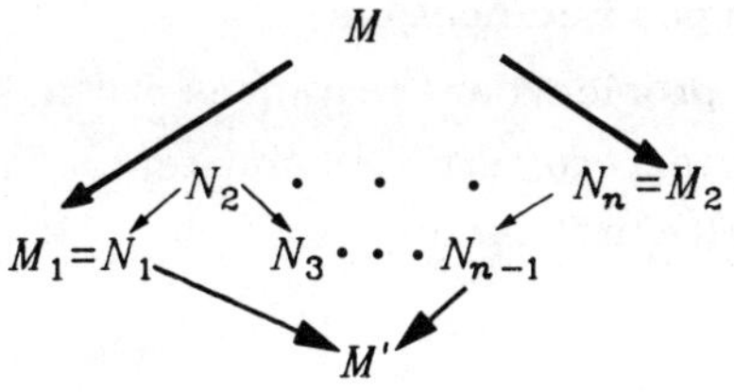

Fig. 2.8

Such a diagram is used to prove the inclusion $=_R \subset =_{R_\infty}$ in the proof of Prop.2.13. It will also be used at the end of chapter 5. An account of other confluence results can be found in a paper by J. Staples [Sta75].

Theorem 2.10 shows that in case of success, we have $=_{R_\infty} \equiv =_{R_k} \equiv =_E$, for some iteration k. Consequently we get a decision procedure for the validity problem in the variety $\mathbf{M}(E)$:

$$\mathbf{M}(E) \models M = N \Leftrightarrow M =_E N \Leftrightarrow R_k(M) = R_k(N).$$

where the last equality is the syntactic one.

Finally, the commutativity axiom or permutations axioms are a stumbling-block for this method: the equation $x+y=y+x$ cannot be introduced as a rule without loosing the termination property of the set of rules. Structural methods have also been developed to handle such cases. In the next section we describe the main one.

2.3 THE ASSOCIATIVE COMMUTATIVE CASE

This section can be omitted in a first reading. Our goal is the analysis of the completion in classical varieties: groups, rings, etc... In this context, the commutativity law appears together with the associativity. As mentioned in the previous chapter, there exists a complete associative-commutative unification algorithm. For these theories the commutativity axiom cannot be orientated into a rule. That is why adequate tools have been developed for this frequent case. As the three fundamental operations on terms are the equality, the match and the unification, we must first resolve them. This is the first step in the study of the completion in associative-commutative theories. Solving these three problems in a general framework simplify greatly the amount of work needed in chapters 3 and 4.

2.3.1 Equivalence, matching and unification

Observe that the first two problems are instances of the last one if we consider the variables of either one or the two terms as constants. For efficiency, this does not authorize us to deal only with unification.

2.3.1.1 Equivalence

The operator domain of our theory splits into two parts: $F = AC \cup G$, AC being the set of associative-commutative operators. The equality or equivalence of M and N in the theory whose operator domain is AC, and whose equations are $f(x,y)=f(y,x)$ and $f(f(x,y),z)=f(x,f(y,z))$ for all f in AC, is noted $M =_{AC} N$.

We now drop the notion of term for AC-terms (which we shall call term by misuse). These AC-terms are *flattened* terms with respect to associative-commutative operators.

Definition 2.15

Let $M \in T(F,V)$, the AC-term $AC(M)$ of M is defined by

—If $M \in V$ then $AC(M) = M$.

—If $M = f M_1 \cdots M_n$ and $f \notin AC$ then $AC(M) = f\ AC(M_1) ... AC(M_n)$.

—If $M = f\ P\ Q$ and $f \in AC$ then $AC(M) = f\ AC(M_1) ... AC(M_n)$

where $(M_1, \ldots, M_n) = Norm(f, P, Q)$

and $Norm(f, N_1, \ldots, N_k) =$ if $N_1 = f N'_1 N'_2$

then $Norm(f, N'_1, N'_2, N_2, \ldots, N_k)$

else $(N_1, Norm(f, N_2, \ldots, N_k))$.

Strictly speaking, AC-terms must be defined with lists, as for example, if f, g are AC-operators, $fxgxyz$ may denote either $f(x,g(x,y,z))$ or $f(x,g(x,y),z)$. However, we will not find such ambiguous cases in definitions and lemmata. It is now easy to decide the AC-equivalence:

Lemma 2.16

Let M,N be two terms, then $M =_{AC} N$ iff

—$M = N$ in V or

—$AC(M) = f M_1 \cdots M_n$, $N = f N_1 \cdots N_n$ and

either $f \in AC$ and $\exists \pi$, a permutation of $[1,...,n]$, such that

$$M_{\pi(i)} =_{AC} N_i,\ i=1...n.$$

or $\quad f \notin AC$ *and* $M_i =_{AC} N_i$, $i=1...n$.

Thus the problem reduces to the computation of the permutation π. The simpler solution sorts the arguments under the function symbol f. These arguments are finite unordered sequences of elements, or ***multisets***, and we need a notion of multiset ordering also used in the last section of the present chapter. Intuitively, a multiset is a collection of copies of elements from a given set E; or formally a subset of $E \times \mathbb{N}$, or a function $M: E \longrightarrow \mathbb{N}$. With this last definition, the vocabulary of sets extends to multisets:

ϕ is the multiset M such that $\forall x \in S$, $M(x)=0$.

$\{x\}$ is the multiset M such that $M(x)=1$ and $\forall y \neq x \in S$, $M(y)=0$.

$x \in M$ iff $M(x) \neq 0$.

$M \subset N$ iff $\forall x \in X$, $M(x) \leq N(x)$.

etc...

Definition 2.17

Let S be a set and < a partial-ordering on S. Let M(S) be the multiset whose elements belong to S, the multiset ordering ≪ defined by (S,<) is the partial-ordering on M(S) such that:

$\forall F,G \in M(S)\quad F \ll G$ *iff* $\exists\ U,V \in M(S)$ *with*

$U \subseteq G$, $U \neq \phi$, $F = (G-U) \cup V$ *and* $\forall v \in V$, $\exists u \in U\quad v<u$.

Intuitively, the multiset F is smaller than the multiset G if, by deleting in F the copies of common elements between F and G, the remaining multiset is such that every copy is smaller than one from G. This multiset ordering is important as shown by the next theorem [Der79a].

Theorem 2.18

The multiset ordering ≪ is noetherian iff the partial-ordering < is noetherian. Moreover, ≪ is total iff < is total.

In this last case, $\ll$ is the lexicographic ordering on the decreasing sequences on S. Now, we work on the multiset of AC-operators arguments. Let $\leq$ be a total order on the set $F \cup V$. We define an AC-order $\leq_{AC}$:

Definition 2.19

Let $M = f\ M_1 \cdots M_n$, $N = f\ N_1 \cdots N_m$ *be two AC-terms, then* $M \leq_{AC} N$ *iff*

either $f \leq g$,

or $f = g \notin AC$, $m = n$ *and* $\exists\, i \leq n$ *such that*

$M_1 = N_1, \ldots, M_{i-1} = N_{i-1}$ *and* $M_i \leq_{AC} N_i$.

or $f = g \in AC$, *and* $\{ M_1, \ldots, M_n \} \ll_{AC} \{ N_1, \ldots, N_n \}$.

Lemma 2.20

Let M,N *be two terms, then* $M =_{AC} N$ *iff* $AC(M) \nless_{AC} AC(N)$ *and* $AC(N) \nless_{AC} AC(M)$.

The proof is straightforward. For a thorough discussion on the implementation of AC-terms see [Hul80a, chap.3]. This definition allows us to work on a normal form of associative-commutative terms $M = f M_1^{p_1} \cdots M_n^{p_n}$, where $M_i^{p_i}$ represents p_i copies of the AC-term M_i, the sequence of M_i being strictly $\leq_{AC}$-decreasing and the heading operator of M_i being distinct from f. Thus we have resolved the associative-commutative equality.

2.3.1.2 Matching

The associative-commutative matching is similar to the unification problem of §2.1. We have two steps: one of decomposition that halts on a AC-operator, the other solving this last case. In the classical matching, there exists at most one matching substitution modulo α-conversion. In the present case, we may find more than one solution. The decomposition is done by the following function in which a unificand is a pair list of terms or AC-terms:

Msimpl(U,σ) =

If $U = \phi$ **Then** (U,σ)

Else { *Let* (M,N) *be a pair in* U; $U := U - \{(M,N)\}$;

  If $M \in V$

  Then { **If** $M \notin \sigma$ **Then** *Msimpl*$(U, \sigma.\langle M,N \rangle)$

    If $\sigma(M) =_{AC} N$ **Then** *Msimpl*(U,σ)

    Else *Fail* }

  Else { $M = f\ M_1 \cdots M_n$

    If $N \in V$ **Then** *Fail*

Else { $N=g\ N_1 \cdots N_m$

If $f \neq g$ **Then** *Fail*

If $f=g \in AC$ **Then** { $(U' \cup \{(M,N)\},\sigma')$

where $(U',\sigma'):=Msimpl(U,\sigma)$ }

Else $Msimpl(U \cup \{(M_1,N_1),\ldots,(M_n,N_n)\}, \sigma)$ }}} ▪

The following step generates, for a pair (M,N) in the unificand U computed by *Msimpl*, the solution set of a linear diophantine equation associated to (M,N). Intuitively, this set enumerates all the possible combinations of the two argument lists of M and N, and obviously only some of these solutions will yield a match. Let $AC(M)=f\,M_1^{i_1} \cdots M_n^{i_n}$ (resp. N) be the $\leq_{AC}$-ordered AC-term associated to M (resp. N). A match from M to N must substitute to each M_i an AC-term $f\,N_1^{j_1} \cdots N_m^{j_m}$. But some of these M_i can be linked by the subpart still computed of the substitution σ. Also, we say that M_i is equivalent to M_j modulo σ, $M_i \equiv_\sigma M_j$, iff there exists a sequence $x_0, \ldots, x_l$ such that

i) $(x_i, x_{i+1}) \in \sigma$

ii) $x_0 = M_i$ (*resp.* M_j).

iii) $x_p = M_j$ (*resp.* M_i).

Then the exact number of distinct M_i in σ is given by a function $Dist([M_1, \ldots, M_n], \phi, \sigma)$:

$Dist([\,], L, \sigma) := L.$

$Dist([A_1, \ldots, A_p], L, \sigma) :=$

If *there exists a pair* (B,k) *in* L *such that* $B \equiv_\sigma A_1$

Then $Dist([A_2, \ldots, A_p], L - \{(B,k)\} \cup \{(B,k+1)\}, \sigma)$

Else $Dist([A_2, \ldots, A_p], L \cup \{(A_1,1)\}, \sigma)$ ▪

This function computes a list of pairs (P_j, p_j), j=1,...,k, such that in the multiset of the A_is, P_j appears exactly p_j times modulo some binds in σ. The reader may observe that this substitution acts as an environment: when we run through the two terms, σ is the store where we keep the bindings. The substitutions $\tau(M_i)=f\,N_1^{k_1} \cdots N_m^{k_m}$ are given by the non-null solutions of the diophantine equation on the free commutative monoid on m generators:

$$\sum_{i=1}^{k} p_i x_i = <j_1, \ldots, j_m> \qquad \text{(E)}$$

Algorithms solving such problems are known, see [Hue78a, Hul80a]. To a solution of (E), a $q \times m$ matrix $S=(s_j^i)$, we associate the unificand

$$U_S = \{ (M'_1, f\ N_1^{s_1^1} \cdots N_m^{s_m^1}), \ \cdots \ (M'_k, f\ N_1^{s_1^k} \cdots N_m^{s_m^k}) \}$$

where the M'_i are the AC-terms computed by *Dist*.

Associative-Commutative Matching Algorithm

Input *Two terms M and N.*

$U_0 := \{(M,N)\}$; $\sigma_0 := \phi$; $i := 0$;

Loop { **If** *Msimpl*$(U_i, \sigma_i) =$ *Fail* **Then** *Fail*

  Else { $(U_{i+1}, \sigma_{i+1}) :=$ *Msimpl*(U_i, σ_i); $i := i+1$ };

  If $U_i = \phi$ **Then** *return* σ_i;

  Let (M_0, N_0) *in* U_i; $(M,N) := (AC(M_0), AC(N_0))$;

  Let S be a basis of solutions for the diophantine equation associated to (M,N);

  If $S = \phi$ **Then** *Fail*

  Else { *Let* U_s *be a unificand associated to the solution s in S;*

    $U_{i+1} := U_s \cup (U_i - \{(M_0, N_0)\})$;

    $\sigma_{i+1} := \sigma_i$; $i := i+1$ }} ■

It is easy to see that this algorithm halts necessarily, and that all possible choices in the *Solv* step enumerate systematically all the AC-matches from M to N.

Let us see an example. With $M = f(x, x, g(y))$ and $N = f(a, a, b, b, g(c))$ where $f \in AC$, and the ordering on $V \cup F$ defined by $a < b < c < x < y < g$, the algorithm starts by computing the solution of

$$2x_1 + x_2 = <2,2,1>$$

which belongs to the set

$$S = \{(<1,0,0>,<0,2,1>), (<1,1,0>,<0,0,1>), (<0,1,0>,<2,0,1>)\}$$

The corresponding unificands are:

$$\begin{cases} U_1 = \{ (x,a),(g(y),f(b,b,g(c))) \} \\ U_2 = \{ (x,f(a,b)),(g(y),g(c)) \} \\ U_3 = \{ (x,b),(g(y),f(a,a,g(c))) \} \end{cases}$$

The next step fails on unificands U_1 and U_3, while U_2 gives the unique solution:

$$\sigma = \{ <x,f(a,b)> , <y,c> \}$$

The reader may check that we have $\sigma(M) =_{AC} N$. One can see that this matching algorithm needs many term manipulations and is expensive. The associative-commutative unification is in fact more complex.

2.3.1.3 Unification

As for the matching, we have two steps of simplification and resolution. We want a complete and minimal set $ECU_{AC}(M,N,W)$ of unifiers of M and N out of the variable set W. This set includes $V(M) \cup V(N)$, and its use will be clear from the following algorithms.

The difference between matching and unification lies in an occurrence test during simplification, detecting failures due to a common variable as for x and $f(x,a)$ for example, and in the linear diophantine equation now of the type $\sum_i a_i x_i = \sum_j b_j y_j$. The occurrence test and the simplification are:

Occur(x,x):=*true*.

Occur(x,c):=*false* **If** $c \in F$, $\alpha(c)=0$.

Occur$(x,f M_1 \cdots M_n)$:=*Occur*$(x,M_1) \vee \cdots \vee$ *Occur*(x,M_n) ∎

Usimpl(U,σ) =

If $U=\phi$ **Then** *return* (U,σ)

Else { *Let* (M_0,N_0) *be a pair in* U; (M,N):=$(\sigma(M_0),\sigma(N_0))$;

  If $M \equiv_\sigma N$ **Then** *Usimpl*$(U - \{(M_0,N_0)\},\sigma)$

  Else { **If** $M \in V$ **Then** { **If** *Occur*(M,N) **Then** *Fail*

      Else *Usimpl*$(U,\sigma \cup \{(M,N)\})$ }

    If $N \in V$ **Then** { **If** *Occur*(N,M) **Then** *Fail*

      Else *Usimpl*$(U,\sigma \cup \{(N,M)\})$ }

Else { $M=f\ M_1 \cdots M_n$ *and* $N=g\ N_1 \cdots N_m$
If $f \neq g$ **Then** *Fail*
If $f=g \in AC$ **Then** { $(U' \cup \{(M,N)\}, \sigma')$
where $(U',\sigma') := Usimpl(U-\{(M_0,N_0)\},\sigma)$ }
Else
$Usimpl(U-\{(M_0,N_0)\} \cup \{(M_1,N_1), \ldots, (M_n,N_n)\},\sigma)$ } ∎

In this simplification, the processed terms are σ instances of U terms. The reason is that we must introduce in σ only variables not yet bound by σ. We now see the diophantine step of the associative-commutative unification.

Let $M=f M_1 \cdots M_n$ and $N=f N_1 \cdots N_m$ be two AC-terms. We first eliminate common variables in argument lists:

$Elim(M,N)$:= **While** $\exists M_j, N_k \in V$ *such that* $M_j \equiv_\sigma N_k$ **Do**
{ $M:=f\ M_1 \cdots M_{j-1} M_{j+1} \cdots M_n$;
$N:=f\ N_1 \cdots N_{k-1} N_{k+1} \cdots N_m$ };
return (M,N) ∎

The next step introduces new variables needed by the expression of all possible decompositions of the argument lists:

Let $\{(M_1,m_1), \ldots, (M_\mu,m_\mu)\} = Dist(M_1, \ldots, M_n, \phi, \sigma)$

and $\{(N_1,n_1), \ldots, (N_\nu,n_\nu)\} = Dist(N_1, \ldots, N_m, \phi, \sigma)$,

where σ is the unification we are building, and let $Y=\{y_1, \ldots, y_\mu, x_1, \ldots, x_\nu\}$ be a set of $\mu+\nu$ variables distinct from those in W. The diophantine equation associated with M and N is $\sum_{i=1}^{\nu} n_i x_i = \sum_{j=1}^{\mu} m_j y_j$. The homogeneous linear associated equation

$$\sum_{i=1}^{\nu} n_i x_i - \sum_{j=1}^{\mu} m_j y_j = 0 \qquad \text{(H)}$$

has its solutions described by the basis of the free commutative monoid of all (H) positive solutions. As for matching, we want only non-null solutions. But now zero is a forbidden value for x_i and y_j, as these variables are associated to distinct AC-terms. This basis subset may be described by a $p \times q$ matrix where $p=\mu+\nu$ and q is the monoid's number of generators. We suppose that an algorithm generates this matrix, for further details, see [Hue78a]. Let $s=(s_i^j)$ be a solution, we associate to it the substitution $\sigma_s(x_i) = f z_1^{s_i^1} \cdots z_q^{s_i^q}$ (resp. y_i) where $z_1^{s_i^1}$ means that the variable z_j appears s_i^j times in the argument list and the variables z_l are distinct from those still used at any time by the unification (this last condition is checked by the set W). Then we have $\sigma_s(f\ x_1 \ldots x_\nu) = \sigma_s(f\ y_1 \ldots y_\mu)$, i.e. $\sigma_s(M) = \sigma_s(N)$. Now, we generate the unificand U_s whose pairs are $(x_i, \sigma_s(x_i))$ (resp. y_j). Simultaneously, we add to the substitution σ the pairs $<x_i, N_i>$ and $<y_j, M_j>$. This is done by the following function *Dioph* where M,N are *AC*-terms, σ is the current substitution and W the set of protected variables, the integer r is the number of non-null solutions of the diophantine equation of M and N:

Dioph(M,N,σ,W) =
If $r=0$ **Then** *Fail*
Else $\{$ $(\{U_1, \ldots, U_r\}, Z, \tau)$
where U_i *is the unificand associated to the ith solution of (H),*
  Z is the set of new variables from the U_i,
  τ *is the substitution* $\{<x_i,N_i>, i=1 \ldots \nu\} \cup \{<y_j,M_j>, j=1 \ldots \mu\}$ $\}$ ∎

We may now give the associative-commutative unification algorithm due to Stickel [Sti81] (a similar algorithm appears in [Liv76]):

Associative-Commutative Unification Algorithm

Input: *Two terms M and N.*
$W_0 := V(M) \cup V(N)$; $U_0 := \{(M,N)\}$; $\sigma_0 := \phi$; $i := 0$;
Loop $\{$ **If** *Usimpl*(U_i, σ_i)=*Fail* **Then** *Fail*
  Else $\{$ $(U_{i+1}, \sigma_i) :=$ *Usimpl*(U_i, σ_i); $W_{i+1} := W_i$; $i := i+1$ $\}$;

  If $U_i = \phi$ **Then** *return* σ_i
  Else $\{$ *Let* (A_0, B_0) *in* U_i; $(A,B) :=$ *Elim*$(AC(A_0), AC(B_0))$;
    If *Dioph*(A,B,σ_i,W_i)=*Fail* **Then** *Fail*
    Else $\{$ $(\{U_1, \ldots, U_r\}, Z, \sigma) :=$ *Dioph*(A,B,σ,W_i);
      Let U *in* $\{U_1, \ldots, U_r\}$;
      $U_{i+1} := [U_i - \{(A_0, B_0)\}] \cup U$;
      $\sigma_{i+1} := \sigma_i \circ \sigma$;

$W_{i+1} := W_i \cup Z$;
$i := i+1$ }}} ■

Theorem 2.21

Let M and N be two terms, if M and N are not AC-unifiable, then the unification stops in failure, else the algorithm halts, and the set of substitutions computed by all choices in the Solve step is a minimal $ECU_{AC}(M,N,V(M)\cup V(N))$.

Stickel [Sti81] showed the correctness in the case where the variables argument of an *AC*-operator did not appear under another *AC*-operator. Fages [Fag84a] proved its correctness and termination in the general case.

2.3.2 Rewriting systems of associative-commutative theories

We obtained algorithms performing the elementary computations of associative-commutative theories: equivalence, matching and unification. Also we may study the rewritings modulo associative-commutative equality. The current theory has its defining equations split in two sets E and R. We suppose that the equations in R are oriented into rules. The two classical properties of rewriting, confluence and termination, must be revisited. The general case (not restricted to AC axioms) has recently being precised by J.P. Jouannaud and H. Kirchner. For a full description of the theory of completions modulo a set of equations, we refer to [Jou84, Kir85].

First the noetherianity, we extend the notion of reduction ordering to E-reduction orderings, to handle the E-equality:

Definition 2.22

A partial-order $\leq$ *on terms, whose strict part* $<$ *is a reduction ordering, is a E-reduction ordering iff* $\forall M,N\ M =_E N \Rightarrow M \leq N$ *and* $N \leq M$

Thus if R satisfies the following property: $\lambda \rightarrow \rho \in R \Rightarrow \rho < \lambda$, then the relation $\xrightarrow{R} . =_E$ is noetherian on *AC*-terms.

Now, we would like a completion algorithm to test the confluence of a reduction modulo associative-commutative equality. To achieve this goal, many relations have been tested. Clearly, we want to simulate the relation induced by R on the quotient $T(F,V)/=_E$. In order to perform a reduction step in this quotient, someone must certainly do many reconfigurations of a term in order to find the good one having a subterm instance of a left member rule. For example, if $f \in AC$, the term

$f(f(x,g(y)),f(a,f(z,b)))$ is reducible by the rule $f(b,g(x)) \rightarrow x$, but the good configuration allowing the reduction must be found. We want to avoid such tricky cases. A first relation closely related to the quotient one is the following $\rightarrow$ $\equiv$ $=_E \cdot \xrightarrow[R]{}$.

Definition 2.23

The rewriting system R is

—E-confluent iff $\rightarrow$ is confluent in $\mathbf{T}(F,V)/=_E$.

—E-noetherian iff $\rightarrow$ is noetherian.

—E-canonical iff $\rightarrow$ is both E-confluent and E-noetherian.

The diagram of the first definition is:

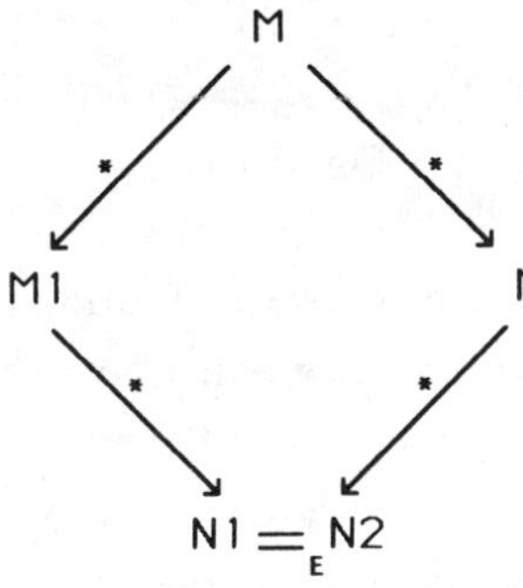

Fig. 2.9

From now on, the rewriting system R is assumed to be E-noetherian, we want to check its E-confluence. The relation $\rightarrow$ being complex by the E-equality step, we introduce a simpler one:

Definition 2.24

For all terms M and N, $M \xrightarrow[R,E]{} N$ iff $\exists\, u \in O(M)$ such that $\exists \lambda \rightarrow \rho \in R$ and a substitution σ with $M/u =_E \sigma(\lambda)$ and $N = M[u \leftarrow \sigma(\rho)]$.

Thus, we now check the associative-commutative equality in the current subterm. As the relation $\rightarrow$ is noetherian, the relation $\xrightarrow[R,E]{}$ terminates, for the latter is included in the former. But we must find conditions under which the second relation is powerful enough to simulate the first one, and we do not forget our goal: the E-confluence of $\rightarrow$. Also, we introduce a new confluence diagram that binds the two relations:

Definition 2.25

The rewriting system R is E-coherent iff

$\forall M,N,\ M \longrightarrow N \ \Rightarrow\ \exists O,P,Q,\ M \underset{R,E}{\Longrightarrow} P,\ P \overset{*}{\longrightarrow} Q\ ,\ N \overset{*}{\longrightarrow} O\ ,\ Q =_E O.$

The corresponding diagram is:

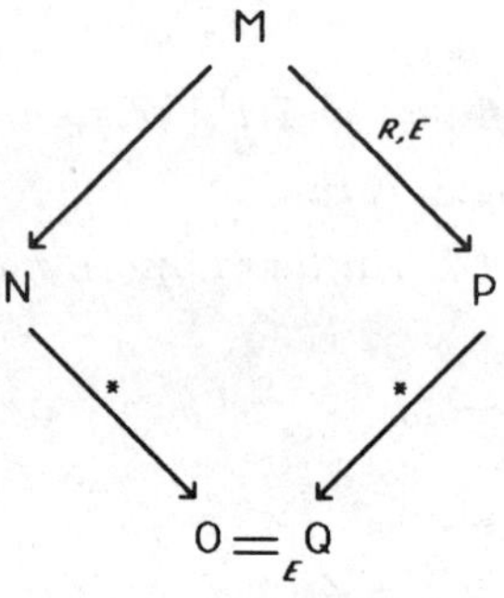

Fig. 2.10

Thus the E-coherence implies that the two reductions are essentially the same. As the last relation is localized, we must now consider critical pairs. The superposition naturally extends to the present case. The difference is that we now E-unify terms, and consequently, more than one critical pair may exist resulting from a given superposition. Thus, we must now speak of a complete set of critical pairs. Moreover, the associative-commutative equation E may also give critical pairs with the rules R:

Definition 2.26

A critical pair of R on E is a critical pair computed by the superposition of a rule $\alpha \longrightarrow \beta$ in R on a pair $(\gamma\ ,\ \delta)$ such that $(\gamma\ ,\ \delta)$ or $(\delta\ ,\ \gamma)$ belongs to E.

It is unnecessary to consider the symmetric case $(\gamma\ ,\ \delta) \in E$ on $\alpha \longrightarrow \beta \in R$ as $\sigma(\alpha[u \longleftarrow \tau(\delta)]) =_E \sigma(\alpha) \underset{R,E}{\Longrightarrow} \sigma(\beta)$, and the critical pair is automatically resolved. This new definition of critical pair is necessary to ensure the E-coherence. Let us assume this condition satisfied, then we have to specify the fact that all R critical pairs are resolved with respect to the relation $\longrightarrow$. This property will be called the E-closure of R. A complete set of R-critical pairs is the union of the complete sets of critical pairs between rules, and between rules and equations.

Definition 2.27

The rewriting system R is E-closed iff there exists a complete set C of R critical pairs such that

$$\forall P,Q \in C\ \exists P_1,Q_1\ P \xrightarrow{*} P_1\ ,\ Q \xrightarrow{*} Q_1\ \&\ P_1 =_E Q_1$$

We thus have the following theorem, which is the associative-commutative version of the Knuth-Bendix theorem:

Theorem 2.28

If the rewriting system R is E-noetherian, E-coherent and E-closed, then it is E-confluent.

This theorem, however, is only a sufficient condition, first established by G.E. Peterson and M.E. Stickel [Pet82], it was extended by G. Huct [Fag83a]. It applies to theories other than the associative-commutative one, provided that, for each $M=N$ in E, $V(M)=V(N)$. The general theory of rewriting modulo a set of equations has been analysed by J.P. Jouannaud and H. Kirchner [Jou84, Kir85]. It uses the assumption of the existence of a E-unification algorithm. Several authors have developed other approaches: J. Pedersen [Ped84] uses substitutions mapping variables to sets of terms; D. Lankford and M. Ballantyne describe completion procedures for theories with associative, commutative or permutative axioms [Lan77a,b,c].

Thus, we must fulfil the three conditions in Thm.2.28. Termination is assumed to be proved by a E-reduction ordering. The E-closure is tested by a new version of the Completion procedure. The last one is achieved by the following result, due to G.E. Peterson and M.E. Stickel [Pet82].

Proposition 2.29

Let R be a rewriting system, the extended rewriting system R_e defined by:

i) $R \subset R_e$.

ii) *If* $\lambda \rightarrow \rho \in R$ *and the leading symbol of* λ *is* $f \in AC$ *then* $f(\lambda,x) \rightarrow f(\rho,x))$, *x a variable, belongs to* R_e.

is E-coherent and the theories $R \cup E$ *and* $R_e \cup E$ *are equal.*

The proof can be found in [Pet81]. Thus, we now have the new completion algorithm, where $(\alpha \rightarrow \beta)_e$ denotes the extension of the rule $\alpha \rightarrow \beta$ according to the previous proposition.

Associative-Commutative Completion Algorithm

Input R : A finite set of equations,

E : The associative-commutative axioms

$<$: A E-reduction ordering,

Super(k,R) : Computes critical pairs between rule k and rules in R whith label less or equal than k, and with the equations from E.

$R_0 := \phi$; The set of rules.

$E_0 := R$; The set of waiting equations.

$i := p := 0$; The step counter and the rule counter.

Loop { **If** $E_i \neq \phi$

Then { *choose a pair* $(M_0, N_0) \in E_i$; *Create_Rule*$(R_i(M_0), R_i(N_0))$ }

Else { **If** *all rules are superposed*

Then *Success*

Else { *choose a (extended) rule with label k not superposed with other rules;*

$E_{i+1} := Super(k,R)$; *mark (extended) rule* k;

$R_{i+1} := R_i$; $i := i+1$ }}}

where Create_Rule(M,N) =

If $M =_E N$

Then { $E_{i+1} := E_i - \{(M_0, N_0)\}$; $R_{i+1} := R_i$; $i := i+1$ }

Else { **If** $M > N$ **Then** { $\alpha := M$, $\beta := N$ }

If $M < N$ **Then** { $\alpha := N$, $\beta := M$ }

Else *Fail*;

Let K be the set of rules in R_i *whose left member is reducible by the rule* $\alpha \rightarrow \beta$ *or its extension* $(\alpha \rightarrow \beta)_e$.

$E_{i+1} := E_i - \{ (\lambda, \rho) \mid k : \lambda \rightarrow \rho \in K \}$;

$R_{i+1} := \{ k : \lambda \rightarrow \rho' \mid k : \lambda \rightarrow \rho \in R_i - K,\ \rho' = (R_i \cup \{\alpha \rightarrow \beta,\ (\alpha \rightarrow \beta)_e\})(\rho) \}$
$\cup \{ p+1 : \alpha \rightarrow \beta,\ (\alpha \rightarrow \beta)_e \}$;

$p := p+1$; $i := i+1$ } ∎

The correction of the algorithm may be proved as in the general case, as all algorithms of equivalence, matching and unification are complete.

2.4 REWRITING ORDERINGS AND CANONICAL SYSTEMS

The previous sections introduced the background of rewriting theory. Now, we present the application of this theory to the classical algebras (monoids and semi-groups, groups, rings modules and algebras, commutative or not). Before this presentation, we need some tools to prove the termination.

2.4.1 Rewriting orderings

First of all, the termination of a rewriting system is an undecidable question. This result was proved by Huet and Lankford [Hue78b], by coding the next move function of a Turing Machine into a set of rules on monadic terms, so that the Turing Machine halts on every instantaneous description iff the rewriting does so on every term. In the same time, they proved that a ground rewriting system was always terminating.

Theorem 2.30

The uniform halting problem for term rewriting systems is undecidable, even for terms restricted to monadic function symbols.
The uniform halting problem for ground rewriting systems is decidable.

The major point for us in this coding lies in the fact that only constant or unary functions are defined on the state set of the Machine. Consequently, for all orderings we may define, we will find counter-examples of rule sets whose noetherianity may not be checked by this ordering. We shall see in the following chapters such counter-examples.

Several orderings have been designed. Let us present three classical ones: first, the Knuth-Bendix ordering [Knu70] based on a weight function on terms; second a recursive path ordering closely related to the term structure; and, finally, an ordering mapping terms on integers. The associative-commutative theories, however, are difficult to handle with orderings: the embedding rules do not respect the term structure, as the terms are flattened. Also, the first two orderings are not subtle enough to reflect this structure manipulation.

2.4.1.1 The Knuth Bendix ordering

The operator domain F is totally ordered by $<_F$, and valued by a weight function $\pi: F \longrightarrow \mathbb{N}$ satisfying the following two conditions:

i) $\alpha(f) = 0 \Rightarrow \pi(f) > 0$,

ii) $\alpha(f) = 1$ and $\pi(f) = 0 \Rightarrow f$ maximal for $<_F$.

For a term M, the integer $n(x,M)$ is the number of distinct occurrences of the variable or function symbol x in M. In order to define the weight of M, we must attribute a common weight to the variables such that the inequality $\pi(\sigma(M)) \geq \pi(M)$ holds for all substitutions σ.

$$\forall x \in V \quad \pi(x) = \min_{\alpha(f)=0} \pi(f)$$

The weight of a term is then $\pi(M) = \sum_{f \in F \cup V} \pi(f) n(f,M)$. Let us extend the ordering $<_F$ in a total one on $F \cup V$ with $f <_F x$ for all $x \in V$ and $f \in F$.

Definition 2.31

Let M and N be two terms, $M > N$ iff

$\forall x \in V(M) \cup V(N)$ $n(x,M) \geq n(x,N)$ *and*

either $\pi(M) > \pi(N)$,

or $\pi(M) = \pi(N)$ *and*

either $M = f(f \ldots (f(x) \ldots)$, $N = x$ *and* f *is maximal for* $<_F$,

or $M = f M_1 \cdots M_n$, $N = g N_1 \cdots N_m$ *and*

either $g <_F f$,

or $g = f$ *and*

$\exists i \leq n$ $M_1 = N_1, \ldots, M_{i-1} = N_{i-1}$ and $M_i > N_i$.

The first condition i) is necessary: rules of the from $c \to M$, c constant, M closed term, may be created. Moreover, if all constants'weight were null, then the variables'weight would be zero. The orientation of rules with variables of the right member not occuring in the left one would be allowed, which makes non-sense. The second condition allows us to forget a function with respect to the weight valuation. This is useful in theories with inverses such as the group one.

Proposition 2.32

A Knuth-Bendix ordering $>$ is a reduction ordering. It satisfies the following properties:

i) for all terms M, $M > M/u$, $\forall u \in O(M) - \{\varepsilon\}$;

ii) it is compatible.

iii) it is total on ground terms.

If R is a rewriting system such that $\forall \lambda \to \rho \in R$, $\lambda > \rho$, then R is noetherian.

See [Knu70] for a proof. The only non-trivial point is the demonstration of the stability: $\sigma(M)>\sigma(N)$ if $M>N$. This is done by induction on the size of M and N. The property i) tells us that a Knuth-Bendix ordering is consistent with the term hierarchy. The property ii) recalls that > is a reduction ordering (cf. Def.2.5). Orderings satisfying i) and ii) are important in proving termination. They are called simplification orderings [Der79a, Der79b].

Theorem 2.33

If R is a rewriting system and $>$ a simplification ordering stable under substitution, then the inclusion $R \subset >$ implies R noetherian.

Now, we introduce another simplification ordering called the *Recursive Path Ordering*, or RPO.

2.4.1.2 The recursive path ordering

This class of orderings (many variants exist) has been introduced by Plaisted and Dershowitz [Pla78a, Pla78b], and uses the multiset orderings introduced in §2.2.

Definition 2.34

Let $>$ be a partial ordering on $F \cup V$, the RPO $>$ defined by $>$ is $M = fM_1 \cdots M_n > N = gN_1 \cdots N_m$ iff

either $f = g$, $\{M_1, \ldots, M_n\} \gg \{N_1, \ldots, N_m\}$ *and* $\{M\} \gg \{N_1, \ldots, N_m\}$

or $f > g$ and $\{M\} \gg \{N_1, \ldots, N_m\}$

or $f > g$ $f \neq g$ *and* $\{M_1, \ldots, M_n\} \gg \{N\}$ *or* $\{M_1, \ldots, M_n\} = \{N\}$

where $\gg$ is the multiset ordering defined by $>$.

Theorem 2.35

Let $>$ be the RPO associated to $>$, then

i) *$>$ is noetherian iff $>$ is noetherian*

ii) *When $>$ is noetherian, $>$ is a reduction ordering.*

This definition being based only on terms a RPO will often be more powerful in proving termination than a Knuth-Bendix ordering. Other orderings have been proposed; we mention the recursive decomposition ordering [Jou82].

2.4.1.3 The integer polynomial interpretation ordering

This method is a special case of what is called the Increasing Interpretation Method in the survey of Huet and Oppen [Hue80b]. The key idea, introduced by Z. Manna and S. Ness [Man70], is to define a function φ from terms in a well-founded set such that $M \rightarrow N$ implies $\varphi(M) > \varphi(N)$ in this set. The first observation is that, when the ground term algebra is not empty, a rewriting system is noetherian iff it is noetherian on the ground term subalgebra. Then we take as well-founded domain the set of natural integers $\mathbb{N}$. This set can be turned in a F-algebra by giving an interpretation of each operator in F. When these interpretations are increasing ones, we get the compatibility condition of Def.2.5: $M>N \Rightarrow P[u \leftarrow M]>P[u \leftarrow N], \forall u \in O(P)$.

We must still prove the stability, $M>N \Rightarrow \sigma(M)>\sigma(N)$. This condition is implied by the stronger one $\forall\ \lambda \rightarrow \rho \in R \ \Rightarrow\ \varphi(\lambda) > \varphi(\rho)$. Thus this last condition suffices to prove the noetherianity of the system when terms are interpreted on $\mathbb{N}$ by increasing functions. This method is frequently untractable, however, as it involves proving integer polynomial inequalities, undecidable in general. Nevertheless, this method sometimes remains the only valid one, we shall use it especially in the case of associative-commutative rewriting systems.

2.4.2 Canonical Systems

Numerous implementations of the completion procedure exist nowadays, the first one being of course written by Knuth and Bendix. Due to their great complexity, very few implementations of associative-commutative completion are available (while some are under implementation). We mention those of Stickel [Sti84], the KB system developed by Hullot and Huet [Fag84b], the REVE system [Les83] at Nancy, partly developed in common with the RRL system of General Electric, described in [Gut84]. Some systems described below were previously found by Knuth and Bendix, while others were hand-checked.

The analysis of the complete sets of rules we now present may be found in Hullot Thesis [Hul80a,b], especially the details of the completion. First of all, let us say that more than one canonical system may exist for a given class of algebras. What is called "the" canonical system of a variety is the one computing the usual irreducible forms in groups or rings. For a given ordering and when the rules are inter-reduced (as in this monograph), this system is unique as shown by Métivier [Mét83].

The semigroups are defined by the associativity axiom of a single binary operator. Any of the two possible orientations gives a canonical system, say

$$\mathbf{A} = \{(x.y).z \rightarrow x.(y.z)\}$$

This orientation results from the RPO, thus proving the termination of the system. The canonical form of a term is $(a_1.(...(a_{n-1}.a_n)...)$ where the a_i are variables. Note that solving the associative equality is not fully satisfactory. Here is an example of Peterson and Stickel: the two rules $a.b \to a$ and $a.x.b \to x.b$, where x is a variable and a,b are constants, cancels all as to the left of the rightmost b using rewriting modulo associativity. However, the general completion procedure will generate infinitely many rules $a.(x.b) \to x.b$, $a.(x.(y.b)) \to x.(y.b) \cdots$ Such cases do not appear when the additional equations do not contain variables, as for presentations of algebras.

The introduction of a unit element 1 for the binary operator defines the variety of monoids. The canonical system is (from now on the associativity rule will always be oriented as in **A**):

$$\mathbf{M}\left\{\begin{array}{lcl} (x.y).z & \to & x.(y.z) \\ x.1 & \to & x \\ 1.x & \to & x \end{array}\right.$$

Once more, the RPO proves the noetherianity of the system, independently of the domain operator ordering. Now, the canonical form of the terms is the semigroup one, as the unit element is always removed from terms by the two last rules, except for the unit element itself, in canonical form.

The following three axioms $x.(y.z)=(x.y).z$, $1.x=x$, $x^{-1}.x=1$ define the variety of groups. Their completion gives the following celebrated canonical system, which appeared first in the paper by Knuth and Bendix [Knu70]:

$$\mathbf{G}\left\{\begin{array}{lcllcl} x.1 & \to & x & 1.x & \to & x \\ x.x^{-1} & \to & 1 & x^{-1}.x & \to & 1 \\ x.(x^{-1}.y) & \to & y & x^{-1}.(x.y) & \to & y \\ 1^{-1} & \to & 1 & & & \\ (x.y).z & \to & x.(y.z) & & & \\ (x.y)^{-1} & \to & y^{-1}.x^{-1} & & & \\ (x^{-1})^{-1} & \to & x & & & \end{array}\right.$$

Its termination is proved by the Knuth-Bendix ordering defined by $^{-1} > . > 1$, and $\pi(^{-1}) = 0$, $\pi(.) = \pi(1) = 1$. Note that the rule $(x.y)^{-1} \to y^{-1}.x^{-1}$ needs the weight of $^{-1}$ to be zero, this is the justification of the special case in the Knuth-Bendix ordering definition. The RPO may also be used, with for example $^{-1} > . > x \equiv y \equiv z > 1$.

For the commutative case of these three varieties, the semigroups have no rules, their definition being handled by the structure of the associative-commutative completion. The monoids have a canonical system reduced to a single rule:

$$\mathbf{M}_c : x+0 \rightarrow x$$

The study of associative-commutative reduction has underlined the need of extended rules in order to insure the coherence. In the present case, this extension is $x+y+0 \rightarrow x+y$, but its left member is reducible by the previous one by the substitution $\sigma(x) = x+y$. Also, the extended rule is redundant. This fact will appear many times, and we will give only the non-redundant extension rules.

The abelian groups are defined by the equations $x+0=x$ and $x+(-x)=0$, the operator + being associative and commutative, the canonical system is [Pet81] (see also [Lan77c]):

$$\mathbf{G}_C \begin{cases} x+0 & \rightarrow x \\ x+(-x) & \rightarrow 0 \\ x+(-y)+y & \rightarrow x \\ -0 & \rightarrow 0 \\ -(-x) & \rightarrow x \\ -(x+y) & \rightarrow (-x)+(-y) \end{cases}$$

The termination is proved by an integer polynomial interpretation:

$$\begin{aligned} \varphi(+) &= \lambda x_1 \cdots x_n . x_1 + \cdots + x_n \\ \varphi(-) &= \lambda x . x^2 \\ \varphi(0) &= 2 \end{aligned}$$

As an example, we have $\varphi(-(x+y)) = 16$ and $\varphi((-x)+(-y)) = 8$. The only rule in $\mathbf{G}_C$ with a non-trivial extension is $(-x)+x \rightarrow 0$. Observe that left members are no longer inter-reduced, due to the presence of extended rules. For example, note that according to Def.2.24, the term $a+(b+(c+(-b)))$ is not reducible by the rule $x+(-x)\rightarrow 0$, but by its extension.

Now, we may present the next canonical systems for the remaining algebraic structures. The previous ones had relatively independent complete presentations, while the commutative presentation $\mathbf{G}_C$ will always be included in the canonical systems. The rings are defined as abelian groups under the operators + and 0, with a multiplicative structure defined by an associative operator ., the two operators + and . being related by the distributive laws:

$$x.(y+z)=x.y+x.z \text{ and } (y+z).x=y.x+z.x$$

This is the usual fundamental structure of rings, over which many variations are possible. Some authors do not suppose the multiplication to be associative; it may possess a unit element and/or be commutative. All these variants, for our purpose, are distinct only in adding or removing some rules from a basic complete set of rules. Thus we give two sets: one for unitary associative rings, and another for unitary associative-commutative rings. The first one is [Hul80a,b]:

$$
R_U \left\{
\begin{array}{lll}
x+0 & \rightarrow x & \\
(-x)+x & \rightarrow 0 & x+(-y)+y \rightarrow x \\
1.x & \rightarrow x & (x.y).z \rightarrow x.(y.z) \\
x.1 & \rightarrow x & x.(y+z) \rightarrow (x.y)+(x.z) \\
-0 & \rightarrow 0 & (x+y).z \rightarrow (x.z)+(y.z) \\
-(-x) & \rightarrow x & -(x+y) \rightarrow (-x)+(-y) \\
x.0 & \rightarrow 0 & x.(-y) \rightarrow -(x.y) \\
0.x & \rightarrow 0 & (-x).y \rightarrow -(x.y)
\end{array}
\right.
$$

The termination follows from an interpretation over integers:

$$
\begin{array}{lll}
\varphi(+) & = & \lambda x_1 \cdots x_n . x_1 + \cdots + x_n + 5 \\
\varphi(.) & = & \lambda x_1 \cdots x_n . x_1 \times \ldots \times x_n \\
\varphi(-) & = & \lambda x . 2 \times (x-1) \\
\varphi(0) & = & 2 \\
\varphi(1) & = & 2
\end{array}
$$

This canonical system defines formally and validates the usual notation for rings elements as a ***polynomial*** over integers one: $\alpha = \sum_i \varepsilon_i a_{i_1} \cdots a_{i_n}$, where ε_i indicates the presence or absence of the inverse operator -, each a_{i_j} is either a variable or a generator of a ring. We do not yet have the usual polynomials, with *scalar coefficients*. These coefficients will be elements α of the base ring.

The commutative rings have a canonical system of ten rules [Pet81]:

$$
R_{CU} \left\{
\begin{array}{lll}
x+0 & \rightarrow x & \\
(-x)+x & \rightarrow 0 & x+(-y)+y \rightarrow x \\
1.x & \rightarrow x & x.(y+z) \rightarrow (x.y)+(x.z) \\
-0 & \rightarrow 0 & -(x+y) \rightarrow (-x)+(-y) \\
-(-x) & \rightarrow x & x.(-y) \rightarrow -(x.y) \\
x.0 & \rightarrow 0 &
\end{array}
\right.
$$

As for rings, many variants of modules can be defined, such as left or right modules, bi-modules over rings, commutative or not. For a detailed analysis, see [Hul80a]. We present the canonical system **Mo** of a bi-module over a unitary associative-commutative ring; greek letters denote ring variables:

$$
\begin{array}{lll}
0+\alpha & \rightarrow & \alpha \\
(-\alpha)+\alpha & \rightarrow & 0 \\
\alpha+(-\beta)+\beta & \rightarrow & \alpha \\
0\times\alpha & \rightarrow & 0
\end{array}
$$

$$
\begin{array}{lll}
1\times\alpha & \rightarrow & \alpha \\
\alpha\times(\beta+\gamma) & \rightarrow & (\alpha\times\beta)+(\alpha\times\gamma) \\
-0 & \rightarrow & 0 \\
-(-\alpha) & \rightarrow & \alpha \\
-(\alpha+\beta) & \rightarrow & (-\alpha)+(-\beta) \\
\alpha\times(-\beta) & \rightarrow & -(\alpha\times\beta) \\
x\oplus O & \rightarrow & \mathrm{x} \\
\alpha.(\beta.x) & \rightarrow & (\alpha\times\beta).x \\
0.x & \rightarrow & 0 \\
\alpha.O & \rightarrow & 0 \\
1.x & \rightarrow & \mathrm{x} \\
(\alpha.x)\oplus(\beta.x) & \rightarrow & (\alpha+\beta).x \\
x\oplus(\alpha.y)\oplus(\beta.y) & \rightarrow & x\oplus((\alpha+\beta).y) \\
\alpha.(x\oplus y) & \rightarrow & (\alpha.x)\oplus(\alpha.y) \\
\sim(x) & \rightarrow & (-1).\mathrm{x} \\
x\oplus(\alpha.x) & \rightarrow & (1+\alpha).x \\
x\oplus y\oplus(\alpha.y) & \rightarrow & x\oplus((1+\alpha).y) \\
x\oplus x & \rightarrow & (1+1).\mathrm{x} \\
x\oplus y\oplus y & \rightarrow & x\oplus((1+1).y)
\end{array}
$$

The canonical form of the terms is thus $\sum_i \alpha_i x_i$ where α_i is the canonical form of a term in the scalar ring (except 1), and x_i is a variable or a generator of the module.

Apart from the classical rule $(-\alpha)+\alpha \rightarrow 0$, two rules, $x\oplus(\alpha.x)\rightarrow(1+\alpha).x$ and $x\oplus x\rightarrow(1+1).x$, have non-redundant extensions.

Finally, the algebras are modules over a ring and they are themselves rings by a multiplicative law $*$. The associative-commutative case gives a system with twenty-six rules. The scalar ring is unitary and associative-commutative. However, we omit the mention of the ten rules for the scalar ring.

$$
\begin{array}{lll}
x\oplus O & \rightarrow & \mathrm{x} \\
\alpha.(\beta.x) & \rightarrow & (\alpha\times\beta).x \\
1.x & \rightarrow & \mathrm{x} \\
(\alpha.x)\oplus(\beta.x) & \rightarrow & (\alpha+\beta).x \\
x\oplus(\alpha.y)\oplus(\beta.y) & \rightarrow & x\oplus((\alpha+\beta).y) \\
\alpha.(x\oplus y) & \rightarrow & (\alpha.x)+(\alpha.y) \\
\sim(x) & \rightarrow & (-1).\mathrm{x} \\
x\oplus(\alpha.x) & \rightarrow & (1+\alpha).x \\
x\oplus y\oplus(\alpha.y) & \rightarrow & x\oplus((1+\alpha).y) \\
0.x & \rightarrow & 0
\end{array}
$$

$$
\begin{array}{lll}
x \oplus x & \rightarrow & (1+1).x \\
x \oplus y \oplus y & \rightarrow & x \oplus ((1+1).y) \\
\alpha.0 & \rightarrow & 0 \\
x * (y \oplus z) & \rightarrow & (x * y) \oplus (x * z) \\
x * (\alpha.y) & \rightarrow & \alpha.(x * y) \\
x * 0 & \rightarrow & 0
\end{array}
$$

And, if the multiplication operator * is not commutative, then we must add the following four rules:

$$
\left\{
\begin{array}{rcl}
0 * x & \rightarrow & 0 \\
(\alpha.x) * y & \rightarrow & \alpha.(x * y) \\
(x * y) * z & \rightarrow & x * (y * z) \\
(x \oplus y) * z & \rightarrow & (x * z) \oplus (y * z)
\end{array}
\right.
$$

For these last systems, we did not give the rewriting orderings. The reader may however convince himself that they are noetherian and compute the classical normal forms.

Finally, we present complete systems for the inverse property [Knu70], defined by a single axiom $x^{-1}.(x.y)=y$:

$$
\left\{
\begin{array}{lcl}
x^{-1}.(x.y) & \rightarrow & y \\
(x^{-1})^{-1}.y & \rightarrow & x.y \\
x.(x^{-1}.y) & \rightarrow & y
\end{array}
\right.
$$

Central groupoids [Eva51], defined by $(x.y).(y.z)=y$:

$$
\left\{
\begin{array}{lcl}
(x.y).(y.z) & \rightarrow & y \\
x.((x.y).z) & \rightarrow & x.y \\
(x.(y.z)).z & \rightarrow & y.z
\end{array}
\right.
$$

Quasigroups [Eva51, Knu70, Hul80b]:

$$
\left\{
\begin{array}{lcl}
x.(x \backslash y) & \rightarrow & y \\
(x/y).y & \rightarrow & x \\
x \backslash (x.y) & \rightarrow & y \\
(x.y)/y & \rightarrow & x \\
(x/y) \backslash x & \rightarrow & y \\
x/(y \backslash x) & \rightarrow & y
\end{array}
\right.
$$

If we add a unit element 1, the following rules together with the previous ones define a complete system for Loops [Eva51, Knu70]:
$1.x \rightarrow x$, $x.1 \rightarrow x$, $1 \backslash x \rightarrow x$, $x/1 \rightarrow x$, $x/x \rightarrow 1$ and $x \backslash x \rightarrow 1$. Termination is trivial

for all these systems.

Free distributive lattices also possess a canonical system [Pet81], they are defined by two associative-commutative operators $\vee, \wedge$, and by the distributivity law and $x\wedge(x\vee y)=x$, $x\vee(x\wedge y)=x$.

$$
\mathrm{L}_D \begin{cases} x\wedge(x\vee y) & \to x \\ x\vee(y\wedge z) & \to (x\vee y)\wedge(x\vee z) \\ x\wedge x & \to x \\ x\vee x & \to x \end{cases}
$$

Termination follows from the interpretation $\varphi(\wedge)=\lambda x_1 \cdots x_n . x_1+\cdots+x_n+1$, $\varphi(\vee)=\lambda x_1 \cdots x_n . x_1\times\cdots\times x_n$.

Note that we obtain a dual complete system by interverting the $\wedge$ and *join* operators. Boolean rings also have a complete system found by J. Hsiang [Hsi82]. They are commutating rings with unit such that the multiplicative operator $\times$ is idempotent ($x\times x=x$) and the additive one + nilpotent ($x+x=0$), the operator + being the exclusive-or.

$$
\begin{cases} x+0 & \to x \\ x+x & \to 0 \\ x\times 1 & \to x \\ x\times x & \to x \\ x\times(y+z) & \to (x\times y)+(x\times z) \\ x\times 0 & \to 0 \\ -x & \to x+1 \end{cases}
$$

From this complete set, we can deduce one for boolean algebras [Fag83a], by eliminating the remaining logical connectives $\wedge$ (or), $\supset$ (implication) and $\equiv$ (equivalence):

$$
\begin{cases} x\wedge y & \to x\times y+x+y \\ x\supset y & \to x\times y+x+1 \\ x\equiv y & \to x+y+1 \end{cases}
$$

Other well-known varieties do not possess a canonical system, we mention Lie rings (the trouble comes from the permutation identity of Lie brackets $[x,[y,z]]\to[[z,y],x]$ from which $[[x,y],[y,x]]\overset{+}{\to}[[x,y],[y,x]]$, it would be worthy to have an equivalent of AC-unification for such algebras, although the word problem for finitely presented Lie algebras is unsolvable [Bok72]); free lattices (an equivalent of normal form is known [Whi41]) and modular lattices (the word problem for free modular lattices is unsolvable [Fre80]). See Pedersen's Thesis [Ped84] for an account of results on decidability in finitely presented algebras. Pedersen seems to be the first to use an infinite recursive set of rules in solving the free word problem for the variety defined by $(x.(x.y)).x=y$ [Ped84, pp52-58]. This set is defined upon the following two sequences of terms:

$$\begin{cases} C_0 & = y \\ C_1 & = x \\ C_2 & = x.(x.y) \\ C_{i+3} & = C_{i+2}.C_i \end{cases} \qquad \begin{cases} D_0 & = x \\ D_1 & = x.(x.x) \\ D_2 & = x \\ D_{i+3} & = D_{i+2}.D_i \end{cases}$$

Therefore, the set of rules is $\{C_{i+2}C_{i+1} \rightarrow C_i \ , \ D_{i+2}D_{i+1} \rightarrow D_i \ , \ i \geq 0\}$, see section 6.3 on Coxeter groups for other examples of infinite sets of rules.

This catalogue ends the introduction to rewriting theory. Completion of finitely presented algebras in some of the above varieties using these complete sets of rules will be detailed in the next two chapters. These complete sets show the importance in the field of elementary algebra of the notion of rules. They provide both a realistic computation tool and a formalization of some normal form results, as defined for example in Cohn [Coh65], where a notion of ***transversal*** is defined (§2.3), which corresponds to the set of irreducible terms as a set of representatives for congruence classes. Cohn also uses a notion of direct move analogous to the reductions, and orderings eliminate infinite direct moves.

However, rewriting has obviously strong limitations. Its inherent ones are first the undecidability of termination. In fact, the termination as expressed throughout this chapter is certainly too restrictive in some cases. We will meet some examples of locally confluent non-noetherian systems that are terminating as the rules are applied in a specified order. Another restriction is the context-free nature of rewritings. We would like to rewrite a subterm only if the term itself satisfies some properties, or according to an environment. In the same way, an intrinsic limitation is that only equational theories can be completed: fields are out of the scope of the theory ($\forall x \neq 0 \ \ x.x^{-1} = 1$), as is the theory of integers with a max operation having a unit element $-\infty$ ($\forall x \neq -\infty, \ x+(-x) = 0$). An attempt to overcome these difficulties is the use of conditional rewriting systems [Kap84]. Another stumbling-block is of course the undecidability of word problems in groups or semigroups, for example. Also other questions have not yet been investigated: with a set of rules as basic decision procedure, are some more general questions easier to answer in this context, such as subalgebra membership, finiteness, isomorphism, substructure membership... To conclude this section, let us say that rewriting systems are obviously not the only way to compute in discrete algebras. We mentioned at the end of chapter one the residually finite and simple algebras, decidable by two complementary semi-algorithms. For simple alternating groups, we failed to obtain complete sets of rules. Always for groups, the Todd-Coxeter method is rather different (see the end of chapter six for a comparison). For permutation groups or matrices algebras, even if complete sets present some interest, they evidently do not provide the simpler means of computation.

3 Monoids and groups

The first two chapters introduced rewriting theory together with a presentation of its main results in the area of equational theories. Now, we show how these results give us completion algorithms for finite presentations of groups, semigroups or monoids.

The original motivation for this specialization of the completion process is twofold. First, completion of a finite presentation in a variety is nothing more than an attempt to solve the word problem for this algebra, which is in fact the first problem to solve in order to compute in a finitely presented algebra. Thus a close study of this problem is valuable. Second, the KB system [Fag84b], developed by J.M. Hullot, which the author uses for his first investigations, is a general purpose one, performing unnecessary computations on this kind of problem.

A study of the canonical forms and rules from chapter 2 allows a specification of the two fundamental operations of matching and unification. This implies a substantial gain in runtime, as observed by the author in writing a LISP system Al-Zebra [LeC84], designed to complete finite presentations of semigroups, monoids, groups and rings, commutative or not. The next two chapters may be viewed as a program synthesis from a general algorithm through finite sets of equations (the complete systems of classical algebras). This synthesis is double: the data-structure of terms is replaced by simpler ones, while the control-structure of the completion is modified according to its behaviour in the variety. The operations of unification and matching, used on a special type of data characterized by a few variables occurring at fixed places, will become (quasi) straightforward. We first need some definitions.

Let G be a finite set of elements called generators (according to Def.1.8). The set of words over G is the set G^* of finite sequences of generators. The empty list, or empty word will be noted 1. The concatenation is an associative law on G^*, and G^* is a representation of the free monoid on G, used in order to introduce the terms in the first chapter. A non-empty word U is

- subword of V iff $\exists W, W' \in G^*$ such that $WUW' = V$ (resp. proper subword iff $WW' \neq 1$).
- prefix of V iff $\exists W \in G^*$ such that $UW = V$ (resp. proper prefix iff $W \neq 1$).

- suffix of V iff $\exists W\in G^*$ such that $WU=V$ (resp. proper prefix iff $W\neq 1$).

If G is a set of generators, G^{-1} is a copy of the set G whose elements are noted a^{-1}, $a\in G$; if $b=a^{-1}\in G$, then b^{-1} denotes the element a from G. If $U=u_1 \cdots u_n$ is a word from $(G \cup G^{-1})^*$, then U^{-1} is the word $u_n^{-1} \cdots u_1^{-1}$ of $(G \cup G^{-1})^*$, with $1^{-1}=1$. A cyclic permutation of the word U is a word V of the form $u_i \cdots u_n u_1 \cdots u_{i-1}$, for $1\le i\le n$.

The length $|U|$ of U, is the number of generators in U, with the convention that $|1|=0$. The set $(G \cup \{I\})^+$ of non-empty sequences on $G\cup\{I\}$, $I\notin G$, with the concatenation will also be used.

3.1 SEMIGROUPS AND MONOIDS

3.1.1 Semigroups

The associativity law of a single binary operator defines the variety of semigroups. We choose as canonical system the following one:

$\mathbf{A} = \{(x.y).z \to x.(y.z)\}$.

According to Def.1.8, a semigroup S is presented by a set of generators G and a set of equations E over closed terms. S is the quotient of the free semigroup on G by the congruence generated by E. We want to solve the word problem for S: do two closed terms represent the same abstract element in S? The completion procedure tries to solve it when given as input the rule $\mathbf{A}$ and the equations over closed terms E. Let us recall that R_∞ is the set of stable rules that are never reduced nor eliminated by the completion procedure (Thm.2.10).

Lemma 3.1

i) The set G^+ is isomorphic to the set of $\mathbf{A}$-canonical terms from $\mathbf{G}(\{.\} \cup G)$.

ii) Let α_1 be a closed term rule in R_∞. There exists an associated rule β_1 in R_∞ with one variable:

$$\alpha_1 : a_1.(\ldots(a_{n-1}.a_n)\ldots) \to b_1.(\ldots(b_{m-1}.b_m)\ldots)$$

$$\beta_1 : a_1.(\ldots(a_{n-1}.(a_n.x)\ldots) \to b_1.(\ldots(b_{m-1}.(b_m.x)\ldots) \quad n,m\in\mathbb{N},\ a_i,b_j\in G,\ x\in V.$$

iii) Superpositions without the associative rule result from two rules

$$\beta_2 : a_1.(\ldots(a_{n-1}.(a_n.x)\ldots) \to b_1.(\ldots(b_{m-1}.(b_m.x)\ldots)$$

$$\alpha_2 : a'_1.(\ldots(a'_{k-1}.a'_k)\ldots) \to b'_1.(\ldots(b'_{l-1}.b'_l)\ldots)$$

such that $\exists i\in\mathbb{N}\ a'_{i+1-j}=a_{n+1-j},\ 1\le j\le i<\min(n,k)$.

Proof. The first proposition asserts the linear structure of irreducible terms, under the bijective correspondence:

$a_1.(a_2 \cdots (a_{n-1}.a_n) \cdots) \Leftrightarrow a_1 a_2 \cdots a_{n-1} a_n$.

Note that this bijection is defined on G^+, as we work in semigroups, without unit element. The other ones detail all possible superpositions. The set R_∞ (from Thm.2.10) is used just to assert the fact that during completion, if the rule α is never deleted, then at some iteration, the rule β must be generated. If rules possess no variables, they cannot be superposed, and a rule of type α or β cannot be superposed on the associative rule **A**. But in ii), the associative rule **A**: $(x.y).z \rightarrow x.(y.z)$ on a closed term one gives a β rule by the substitution $\sigma=\{<x,a_1>,<y,a_2.(\cdots(a_{n-1}.a_n)\cdots)>\}$, after **A**-normalization of the critical pair. In iii) a α rule on a β one gives a α rule, the substitution is then $\sigma=\{<x,a'_{i+1}.(\cdots(a'_{n-1}.a'_n)\cdots)>\}$. There are no other legal superpositions (remember that we superpose on a non-variable subterm) ▪

The variables of β-like rules will be called **terminal** variables. To every α-rule, or **closed** rule, will be associated such a β rule, the latter being called **extended** rule by analogy with the associative-commutative case.

Thus, we get two consequences. First, the data-structure of words is better in this case than the term one; second, all β-rules are redundant. So that we get a new algorithm, its control structure is the completion one. The following list gives the connection between old and new keywords:

— A term is now a word.

— An occurrence is an index.

— The subterm at occurrence i is the suffix $u_i \cdots u_n$ of $u_1 \cdots u_n$.

— The word U matches the word V iff U is a prefix of V.

— The rule $U \rightarrow V$ superposes on $P \rightarrow Q$ iff $\exists A,B,C$ such that $B \neq 1$, $AB=P$ and $BC=U$. The critical pairs are then (QC,AV) for all the possible choices of B.

— The word W reduces in W' iff $\exists U \rightarrow V$, A,B such that $W=AUB$ and $W'=AVB$.

A crucial point here is to observe that we no longer need the full power of unification. This fact will be constant in chapters 3 and 4. Because the variables are all terminal in this case, we just need matching. Thus, if $R(V)$ is the irreducible form of the word V under the canonical word system R, then we have from Thm.2.10:

Corollary 3.2

If (G,E) is a finite presentation of a semigroup S, and R is a canonical system for S, then the subset Σ of G^+ whose elements are all R-irreducible words, with the operator $\times$ defined by

$$\forall\ W,W' \in \Sigma \quad W \times W' = R(WW').$$

is a semigroup isomorphic to S.

Of course, such a finite set of rules does not always exist, the word problem for semigroups being unsolvable [Nov55]. Tseiten [Tse56] has given the following presentation with undecidable word problem:
$S=(a,b,c,d,e;ac=ca,ad=da,bc=cb,bd=db,eca=ae,edb=be,abac=abace)$.

The variety of monoids has the following canonical system :

$$\mathbf{M}\begin{cases}(x.y).z \rightarrow x.(y.z)\\ x.1 \rightarrow x\\ 1.x \rightarrow x\end{cases}$$

The only difference between semigroups and monoids is that the reductions must remove the constant 1 from words, as done by the last two rules in **M**. As superpositions are performed on a non-variable subterm and rules are inter-reduced, these two rules cannot give any critical pair. The sets of rules on words are a generalization of Thue Systems for which only length reducing replacements are allowed. Many results can be found about such systems in [Coc76, Boo82a, Boo82b, Bau84, O'Du83], especially in connection with language and recursion theory. This research area is presently very active (see R. Book's survey [Boo85]). We now present some of these results.

3.1.2 Some results on Thue systems

In fact, Thue systems allow only length-decreasing rules. They may include a set of equations between words of equal length. We did not make such restrictions on the word rewriting rules; they may increase the length. The equations of a Thue system may be used in any sense, so that we reduce modulo a set of word equations.

Some decidability results have been obtained for monadic and special Thue systems.
- A Thue system T is monadic iff $(u,v)\subset T \Rightarrow v\in G\cup\{1\}$.
- A Thue system T is special iff $(u,v)\in T \Rightarrow v=1$.
- A word W is irreducible iff it is minimal in length in its congruence class.

Cochet and Nivat have proved that if the Thue system T is finite, special and confluent, then every congruence class is a non-ambiguous context-free language. Book [Boo82b] gave a more precise result: if T is finite, monadic and confluent, then if R is a regular language, the union over the irreducible words in R of the congruence classes is a deterministic context-free language. And for special Thue systems, Book et al [Boo82a] established the following result:

If the Thue system T is confluent, special and if the set of words M such that there exist N with $(M,N) \in T$ is a regular language, then the union over a regular language R of the congruence classes is a context-free language.

Here are some decidability results [Boo83]:

If M_T is the monoid defined by the congruence associated to a Thue system T, then the following problems are decidable:

— M_T is free.

— M_T is a group.

— If $H \subset G^*$, the submonoid generated by H is free.

— If $H \subset G^*$, the submonoid generated by H is a group.

This contrasts with non-monadic Thue systems, for which the similar problems (left or right-divisibility, submonoid membership...) are undecidable [Ott84a].

After this small overview of Thue systems, we collect some elementary results about the languages of the rules or of the canonical forms:

Theorem 3.3

i) If S is a finite canonical word rewriting system, then the set of canonical forms is a regular language.

ii) There exists a finite monoid presentation on which the completion algorithm generates an infinite canonical system whose left members are a recursively enumerable non-recursive set.

iii) There exists a finite monoid presentation with non-context-free congruence classes.

Proof For a rule $k:\lambda_i \to \rho_i$, the set of k-reducible words is the regular language $L_i = G^* \lambda_i G^*$. Thus the reducible words belong to $\bigcup_{i \leq n} L_i$, which is regular when n is finite. Its complementary in G^* is also regular. Thus the canonical forms are expressible by a regular expression. (cf. chap. 6 for examples of such descriptions, they are concise in practical cases).

The second statement is a negative answer for the infinite case. It is a direct consequence of the existence of a finitely presented group with an unsolvable word problem [Nov55]. Such a group is transformed into a monoid presentation by adding the equations $aa^{-1}=1$, $a^{-1}a=1$, $a \in G$. With a lexicographic ordering, the completion does not halt and, consequently, generates an infinite canonical set of rules R_∞ (cf. Thm.2.10). But, as the word problem is unsolvable, the reduction is no longer an algorithm, that is, redex searching in a word is not recursive: the set of left members is recursively enumerable not recursive.

For the last proposition, the reader may check that we have in the monoid defined by the two equations $abc = ab$ and $bbc = cb$,
$\{ab^{2^n+1}c^n \mid n \geq 0\} = [abb] \cap a^* b^* c^*$.
Thus the congruence class of abb is not context-free as the intersection of a regular and a context-free language is context-free (example from [Boo82b]) ∎
By i) the set of normal forms is regular. Therefore standard techniques can be used to decide finiteness or infiniteness of the monoid defined by a canonical set of rules (e.g. Thm3.7, p.63 of [Hop79]).

We now turn to some complexity results:

- If R is a finite non-length increasing canonical system, then there exists a linear time algorithm computing the R-canonical form of a word [Boo82a].
- There exists a polynomial time decision algorithm to test whether or not a finite and noetherian system R is confluent [Boo81].

The first result above is illustrated in chapter 6. Also, sufficiently rich complexity classes can be coded by word rewrite systems. Bauer [Bau81] has proved that for any such class C, there exists a canonical system such that the word problem restricted to a single generator is exactly of complexity C (see also [Hor83, Ott84]).

3.1.3 Termination

The first fact about termination of word rewriting systems is that it is an undecidable problem, the proof being exactly the same as in Huet and Lankford on the general case [Hue78b]. However, many systems will fall under the scope of orderings presented in chapter 2. We specialize these orderings to words. The notions of compatibility and stability of a reduction ordering merge in the following property:

$$\forall U, V, M, N, \ U > V \Rightarrow MUN > MVN. \qquad \text{(P)}$$

Thus a reduction ordering for words is noetherian and satisfies (P).

The most natural such ordering, a partial one, is the length of the words ($M < N$ iff $|M| < |N|$ in the set of natural integers). Then we have the lexicographical orderings: if G is totally ordered by <, then this ordering $<_{Lex}$ is defined by:

$M <_{Lex} N$ iff Either $|M| < |N|$,
Or $|M| = |N|$ and if $M = m_1 \cdots m_n$, $N = n_1 \cdots n_m$ where m_i and n_j belong to G, then there exists $k \leq \min(m, n)$ with $m_j = n_j$, $0 < j < k$ and $m_k < n_k$.

Knuth-Bendix orderings are a generalization of Lex orderings. The set $G \cup \{.\}$ (where . is the concatenation) is totally ordered by >, and is assigned a weight function

π: $G \longrightarrow \mathbb{N}$ such that $\pi(a)>0$ whenever $a \in G$. The constant k is the weight of the concatenation. Thus the weight of the word $W=w_1 \cdots w_n$ is $\pi(W)=\sum_i \pi(w_i)+(n-1)k$.

Definition 3.4

The Knuth-Bendix ordering on words is defined by

$M>_{KB} N$ *iff Either* $\pi(M)>\pi(N)$,

Or $\pi(M)=\pi(N)$ *and*

If $|M|>1$, $|N|>1$ *then* $\exists i$ *with* $m_i=n_i$, *and* $m_j<n_j$, $0<j<i$,

If $|M|>1$, $|N|=1$, *then* $.>N$,

If $|N|>1$, $|M|=1$, *then* $M>.$,

If $|M|=|N|=1$ *then* $M>N$ *in* G.

The RPO defined in §2.4.1 also provides a useful ordering on words.

Definition 3.5

Let $G \cup \{.\}$ *be partially ordered by* $>$, $M=g_1 \cdots g_m$ *and* $N=g'_1 \cdots g'_n$ *are two words with length resp.* m *and* n. *The RPO ordering* $>_{RPO}$ *on* $>$ *is defined by*:

$M>_{RPO} N$ *iff*

Either $n>1$ *and*

If $m>1$ *then* $\{g_1, \ldots, g_m\} \gg_{RPO} \{g'_1, \ldots, g'_n\}$ *and* $M>_{RPO} g'_i$, $i=1,\ldots,n$;

If $m=1$ *then* $.>N$ *or* $.>N$ *does not hold but* $\exists i$ *such that* $g_i>_{RPO} N$ *or* $g_i=N$;

Or $n=1$ *and*

If $m=1$ *then* $M>N$;

If $m>1$ *then* $M>.$ *and* $\{M\} \gg_{RPO} \{g'_1, \ldots, g'_n\}$;

where $\gg_{RPO}$ *is the multiset ordering defined by* $>_{RPO}$.

In this definition, the ordering $\gg_{RPO}$ may be replaced by a lexicographic or inverse lexicographic ordering on the argument vector. The RPO ordering is especially interesting when one wishes to eliminate a generator from the rules, by declaring it maximal for the $G \cup \{.\}$ ordering. The unit element is supposed to be less than all other words.

We will see at the end of the next section some heuristic techniques for proving the termination of a rewriting system out of the scope of the previous standard orderings.

3.2 GROUPS

The study of non-abelian groups is done in two steps. The completion of groups is analyzed as in the previous cases. But a new fact appears: the rules are **symmetrized** or **normalized**. This operation is closely related to the canonical system for groups. More precisely, in a theory with canonical rewriting system, some rules of this system may be superposed on the oriented equations of a presentation. The superpositions may be guided to create a **symmetrized** presentation. The word symmetrized comes from small cancellation theory (cf. chapter 5). In a symmetrized presentation, rule members are balanced, and all critical pairs between the canonical system of the variety and the symmetrized one are resolved.

The second part of the study is an analysis of this symmetrization process, reported in Chapter 5, as symmetrization is in fact a slight refinement of Dehn algorithm. Let G be a set of generators, G^{-1} is a copy of G whose elements are noted $a^{-1}, a \in G$. If $b = a^{-1} \in G^{-1}$ then b^{-1} is the generator a. If $U = u_1 \cdots u_n \in (G \cup G^{-1})^*$, $u_i \in G \cup G^{-1}$, then $U^{-1} = u_n^{-1} \cdots u_1^{-1}$, with $1^{-1} = 1$.

3.2.1 Completion

The celebrated canonical system for groups is the following one :

$$\mathbf{F} \left\{ \begin{array}{llll} 1: x.1 \rightarrow x & & 2: 1.x \rightarrow x \\ 3: x.x^{-1} \rightarrow 1 & & 4: x^{-1}.x \rightarrow 1 \\ 5: x.(x^{-1}.y) \rightarrow y & & 6: x^{-1}.(x.y) \rightarrow y \\ 7: 1^{-1} \rightarrow 1 & & \\ 8: (x.y).z \rightarrow x.(y.z) & & \\ 9: (x.y)^{-1} \rightarrow y^{-1}.x^{-1} & & \\ 10: (x^{-1})^{-1} \rightarrow x & & \end{array} \right.$$

By rules **F**3, **F**4, **F**5 and **F**6, not all words of $(G \cup G^{-1})^*$ are **F**-irreducible. Also, we need the following function, which computes the **F**-canonical forms of words:

Let F(W) = **Case** W of $1V$, $aa^{-1}V$ or $a^{-1}aV$ **Then** F(V);
aV **Then Case** F(V) of 1 **Then** a;
$a^{-1}U$ **Then** U;
Otherwise aF(V);
Otherwise W ∎

This function is now used to define the group reduction :

$$W \rightarrow W' \text{ iff } \exists\, U \rightarrow V \text{ a word rule, } A, B \text{ such that } W = AUB \text{ and } W' = \mathrm{F}(AVB).$$

The function F computes the well-known canonical form in free groups, i.e. the reduction generated by the rules $aa^{-1} \rightarrow 1$, $a \in G \cup G^{-1}$, or by the canonical system **F**.

Of course, we could merge these rules with those generated by the completion of a given group, but the previous reduction runs faster, and this fact is essential in an implementation (cf. chapter 6).

We then have the equivalent of lemma 3.1, whose details are left out. The point is that new critical pairs are computed between closed term rules and **F** ones. Expressed as words they give the following:

Lemma 3.6

To the rule $a_1 \cdots a_n \to b_1 \cdots b_m,\ a_i, b_j \in G \cup G^{-1}$ *the completion associates the following critical pairs :*

$$(a_1 \cdots a_{n-1}\ ,\ b_1 \cdots b_m a_n^{-1}), (a_2 \cdots a_n\ ,\ a_1^{-1} b_1 \cdots b_m),$$

$$(a_n^{-1} \cdots a_1^{-1}\ ,\ b_m^{-1} \cdots b_1^{-1}).$$

Proof. We omit a first case that is analyzed by the lemma 3.1 stating that the associative rule superposes on every word rule. Let us detail the substitutions for the remaining cases (to words rules are associated term rules as for lemma 3.1).

Rule **F9** $(x.y)^{-1} \to y^{-1}.x^{-1}$, $\sigma(x)=a_1$ and $\sigma(y)=a_2.(\cdots(a_{n-1}.a_n)\cdots)$.

Rules **F5**,**F6** If $a_1 \in G$, the rule **F6**: $x^{-1}.(x.y) \to y$ is superposed with $\sigma(x)=a_1$ and $\sigma(y)=a_2.(\cdots(a_{n-1}.a_n)\cdots)$;

If a_1 is of the form $(a)^{-1}$, then the rule **F5**: $x.(x^{-1}.y) \to y$ is superposed in the same way;

Finally, it may occur, with term rules in place of word ones, that $a_n = b_m \in V$ and $a_{n-1} \in G$, then the rule **F5** with the substitution $\sigma(a_n)=a_{n-1}^{-1}.y$ and $\sigma(x)=a_{n-1}$ gives the last pair (resp. **F6** if $a_{n-1} \in G^{-1}$) ∎

In fact, the behaviour of the two completion algorithms, the first on terms and the second on words, is not exactly the same. In the latter case we will obviously produce the three critical pairs in one step, while in the former, the superposition with the associative law must occur before the computation of the last pair. However, there always exists an iteration such that in both algorithms all the equivalent critical pairs are produced - that is all we need.

As the three critical pairs in lemma 3.6 are closed to the group variety, we may call them **canonical** (or **normal**) pairs for brevity. Now, the group completion is completely defined. The control structure is modified. We must add one step to compute the canonical pairs of the new rules. We can observe that these canonical pairs compute in fact the elementary operations of the group theory:

$$aU=V \Rightarrow U=a^{-1}V \quad Ua=V \Rightarrow U=Va^{-1} \quad U=V \Rightarrow U^{-1}=V^{-1}$$

and $aU=aV \Rightarrow U=V$ after **F**-normalization of the critical pair $(U, a^{-1}aV)$. This **F**-reduction is also called the free cancellation.

A good heuristic is to give the canonical pairs the highest priority in the set E of waiting equations, and to generate them between the first two steps. On words in **F**-normal form, the matching and unification are the monoid ones. We get the group completion algorithm:

Group Completion Algorithm

Input E : A finite set of equations,

$<$: A reduction ordering,

$R_0:=\phi$; The set of rules.

$E_0:=E$; The set of waiting equations.

$RR_0:=\phi$;The set of deleted rules.

$L_0:=\phi$; The set of rules whose critical pairs are not computed.

$LL_0:=\phi$;The set of rules whose critical pairs are computed.

$C_0:=\phi$; The set of canonical pairs.

$S_0:=\phi$; The set of rules whose canonical pairs are not computed.

$i:=p:=0$;The step counter and the rules counter.

Loop { **If** $RR_i \neq \phi$ **Then** {*choose a pair* $(M_0,N_0)\in RR_i$; $RR_{i+1}:=RR_i - \{(M_0,N_0)\}$}

If $C_i \neq \phi$ **Then** {*choose a pair* $(M_0,N_0)\in C_i$; $C_{i+1}:=C_i - \{(M_0,N_0)\}$}

If $E_i \neq \phi$ **Then** {*choose a pair* $(M_0,N_0)\in E_i$; $E_{i+1}:=E_i - \{(M_0,N_0)\}$};

If $RR_i \cup C_i \cup E_i \neq \phi$ **Then** *Create_Rule*$(R_i(M_0), R_i(N_0))$

Else { **If** $S_i=\phi$ **Then** {

If $L_i=\phi$ **Then** *Success*

Else { *Let k be a rule in* L_i;

$L_{i+1}:=L_i - \{k\}$;

Let E_{i+1} *be the set of critical pairs between k and rules from* LL_i;

$LL_{i+1}:=LL_i \cup \{k\}$; $i:=i+1$ }}

Else { *choose a rule k in* S_i;

$S_{i+1}:=S_i - \{ k \}$;

Let C_i *be the set of canonical pairs of rule k*; $i:=i+1$ }}}

where Create_Rule(M,N) =

If $M=N$ **Then** $i:=i+1$

Else { **If** $M>N$ **Then** { $\alpha:=M$, $\beta:=N$ }

If $M<N$ **Then** { $\alpha:=N$, $\beta:=M$ }

Else *Fail;*

$p := p+1;$

$K := \{k \mid k: \lambda \rightarrow \rho \in R_i$ *and* λ *reducible by* $p: \alpha \rightarrow \beta\};$

$RR_{i+1} := \{(\lambda, \rho) \mid k: \lambda \rightarrow \rho \in R_i,\ k \in K\} \cup RR_i;$

$R_{i+1} := \{k: \lambda \rightarrow \rho' \mid k: \lambda \rightarrow \rho \in R_i,\ k \not\in RR_{i+1},\ \rho' = (R_i \cup \{p: \alpha \rightarrow \beta\})(\rho)\}$
$\qquad \cup \{p: \alpha \rightarrow \beta\};$

$S_{i+1} := S_i \cup \{p\};\ I_{i+1} := I_i \cup \{p\};\ i := i+1$ ■

This version of the completion is detailed so that it corresponds closely to the one we implemented in our LISP system Al-Zebra [LeC84]. The only significant difference is that the reduction of a selected pair in order to create a new rule is replaced by a function called parse_theory. In an implementation of specialized completion procedures, this is needed in order to be coherent with the remaining varieties of polynomials, where symmetrization is more complex and acts not only on normal pairs, but also on the form of the rules. Our way of specialization corresponds to the Def.1.8, where a finitely presented algebra A is defined as the quotient of $B = G(F \cup C)/=_E$ by the congruence $=_R$. The first quotient is computed by the canonical system of the variety, while the present chapter details an efficient method for computing in the second quotient.

The reader can now think to the corollary 3.2 versus 3.4, with the inverse and product operators:

$$\text{If } W \in \Sigma \text{ then } W^{\sim 1} = R(W^{-1}). \qquad \text{(I)}$$

$$\text{If } W, W' \in \Sigma \text{ then } W \times W' = R(WW'). \qquad \text{(P)}$$

Lemma 3.7

Let $\mathbf{G} = (G, E)$ *be a finite group presentation, and* R *a canonical system associated to* E*, then the subset* Σ *of* $(G \cup G^{-1})^*$ *of all* R*-irreducible words with the operations* $^{\sim 1}$ *and* $\times$ *defined by equations* (I) *and* (P) *is a group isomorphic to* **G**.

In fact, the **F**-reduction may be forgotten when we extend R by $R_G = R \cup \{a.1 \rightarrow a,\ 1.a \rightarrow a,\ a.a^{-1} \rightarrow 1,\ a^{-1}.a \rightarrow 1 \mid a \in G\}$, acting on the semigroup $(G \cup G^{-1} \cup \{1\})^*$. In this case, we allow R-reductions on words not in **F**-normal form, this relation refines the previous one. We shall need it in Chapter 5.

If we delete the superposition step in completion, we obtain the symmetrization algorithm, whose name will become clearer in Chapter 6. As in the monoid case, the group word problem is unsolvable [Nov55, Sho67], so the completion may not terminate. However, it halts on finite groups, as first observed by Métivier [Mét83]:

Proposition 3.8

Given a presentation (G,E) of a finite group **G** *and a finite reduction ordering, the completion algorithm always halts in success. Moreover, the number of rules satisfies the inequality :*

$$|R| \leq 2.|G|.|\mathbf{G}|.$$

Proof. Suppose that the completion does not halt. By Thm.2.10; it computes an infinite canonical system R_∞. As **G** is finite and the rules right members are R_∞-irreducible, there exists a word W, right member of infinitely many rules. Let $\Sigma=(W_i)_{i\in\mathbb{N}}$ be the sequence of associated left members. As rules are inter-reduced, W_i is not subword of W_j, $j\neq i$. Moreover, all proper subwords of W_i are R_∞-irreducible. Therefore, we can extract from Σ a subsequence Σ' strictly length increasing. Let $W_i=a_i V_i$, $W_i\in\Sigma'$, $a_i\in G\cup G^{-1}$, the set of all V_i is an infinite set of R_∞-irreducible words, contradicting the finiteness of **G**. The algorithm must halt. We now give an upper bound to $|R|$, the number of rules. Let $(\rho_k)_{k=1,\ldots,n}$ be an enumeration of right members, $\rho_k\neq\rho_l$ if $k\neq l$, then $n\leq|\mathbf{G}|$. If $L_k=\{\lambda_i \mid \lambda_i\to\rho_k\in R\}$, then we have $\sum_k |L_k|=|R|$.

Let $M_k=\{\mu_i \mid \exists a_i\in G\cup G^{-1} \text{ s.t. } a_i\mu_i=\lambda_i\in L_k\}$, we may suppose $\mu_i\neq\mu_j$ if $i\neq j$. Otherwise $a_i\mu_i=_{\mathbf{G}}a_j\mu_i=_{\mathbf{G}}\rho_k$ implies $a_i=_{\mathbf{G}}a_j$ while $a_i\neq a_j$ as the rules are inter-reduced. But this means that a generator was redundant. Eliminating such cases, we get $|L_k|=|M_k|$, as the words in M_k are G-irreducible, we have $|M_k|\leq|\mathbf{G}|$. Thus,

$$|R|=\sum_{k=1}^{n}|L_k|\leq\sum_{k=1}^{n}|\mathbf{G}|\leq|\mathbf{G}|^2.$$

And we get $|R|\leq|\mathbf{G}|^2$. But this upper bound may be improved. Fix one $\mu_i\in M_k$, then μ_i appears at most in $|G\cup G^{-1}|=2|G|$ sets M_l. Otherwise two distinct rules would have the same left member. As the number of μ_i is bounded by $|\mathbf{G}|$, we have $\sum_k|M_k|\leq|G\cup G^{-1}|.|\mathbf{G}|$. In other words, $|R|\leq 2|G|.|\mathbf{G}|$ ∎

As in practice the number of generators is very small, this inequality tells us that we get the multiplication table of the group **G** in $O(|\mathbf{G}|)$ rules. However, this bound is still large as one can see in chapter 6. Of course, this result applies to the monoids and the semigroups. Now, the question of complexity may be asked about the completion of finite groups. Our experience shows that complex situations exist: symmetric groups S_n have presentations whose completion gives a set of rules whose cardinality is in $O(|S_n|) \sim n!$. Moreover, we did not succeed in completing the alternating groups A_n. At the end of chapter 6, a comparison between the present method and the Todd-Coxeter coset enumeration method can be found.

The previous result gives an upper bound for the number of rules in the final set of rules, not on the number of intermediate rules generated. For example, in Métivier's thesis [Mét83] 68 rules are generated to prove that the presentation $(a,b,c,d;ac=1,ca=1,bd=1,db=1,abbbcdd=1,abccb=1)$ defines the trivial group. Indeed, presentations of the trivial group show that the bound of Prop.3.8 cannot be improved: for our completion procedure generate exactly $2.|G|.|\mathbf{G}|$ on such presentations. The completion provides a semi-decision procedure for the question of whether a presentation defines the trivial group. This question was proved undecidable by Rabin [Rab58], following ideas of Markov [Mar51]. Also undecidable are the questions [Rab58]: given a presentation P, is the group presented by P cyclic, finite, free, commutative? Note by the observation following Thm.3.3 that we also have a semi-decision procedure for the finiteness of groups.

For infinite groups, we would like some criteria restricting the number of superpositions. In Chapter 5, we show that the symmetrization is in some important cases powerful enough to solve the word problem. Of course, the results about Thue systems remain true for group complete systems.

3.2.2 A detailed example

Before examining the commutative case for both monoids and groups, let us see a very simple example of group completion, and then some heuristics to prove termination of rewriting systems. We are going to complete the group of the equilateral triangle, defined by the presentation $G=(S,T;S^2=T^2=(ST)^3=1)$:

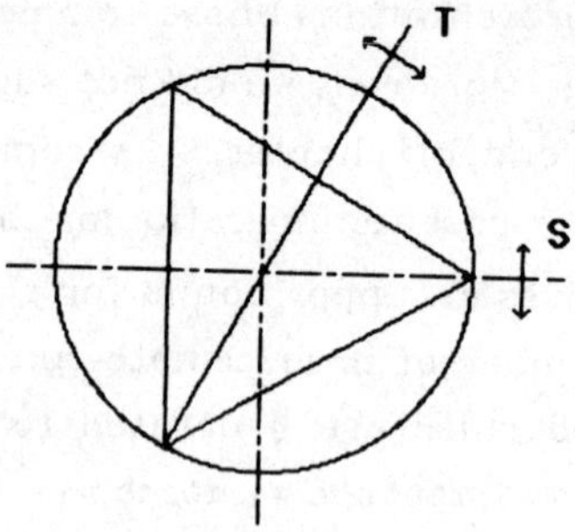

Fig. 3.1

S and T are two reflections whose axis angle is $\pi/3$, their product is a rotation of center O and of angle $2\pi/3$. The algorithm first orientates the three defining equations:

R_1: $SS \rightarrow 1$

R_2: $TT \rightarrow 1$

R_3: $STSTST \rightarrow 1$

Afterwards the symmetrization process starts. The first two rules are simpler than the last one, so they are first selected, creating two new rules;

R_4: $S^{-1} \rightarrow S$

R_5: $T^{-1} \rightarrow T$

These two rules cancel the waiting equations in N_i $(T^{-1}T^{-1},1)$ and $(S^{-1}S^{-1},1)$. Then rule R_3 is symmetrized:

R_6: $TSTST \rightarrow S$

This last rule deletes R_3. The waiting equation created by R_3 is not reintroduced as $STSTST$ reduces to 1 by R_6 and R_1. The symmetrization goes on with the rule R_6 to end at the following system:

$$\begin{cases} SS \rightarrow 1 \\ S^{-1} \rightarrow S \\ TT \rightarrow 1 \\ T^{-1} \rightarrow T \\ TST \rightarrow STS \end{cases}$$

Then the superposition process begins. But no new rule is created as all the critical pairs are resolved. For example, between the third and fifth rules, the word *TTST* gives the pair (*ST*,*TSTS*), and *TSTS*$\rightarrow$*STSS*$\rightarrow$*ST*, so that the previous system is declared complete (its termination is trivial).

This complction clearly shows an iteration on three steps:

1) Orientation of all waiting equations whatever their origin is.
2) Computation of normal pairs.
3) After normalization of all rules, computation of critical pairs.

Chapter 6 is a catalogue of complctc rewriting systems for a number of usual finitely presented groups.

The termination proof techniques used for the monoids apply to the groups. We point out that the existence of the system R^G associated with R leads to two alternatives: the first one considers the elements of G^{-1} as new generators. Thus we may assign them a weight for the Knuth-Bendix ordering, or order them together with the defining generators in a rpo; the other one considers the inverse as an operator. This last possibility is interesting when the generators are cyclic, for in this case the inverses may be eliminated with the following rules, if the inverse operator is declared as the greatest one:

$$\begin{cases} a^n \rightarrow 1 \\ a^{-1} \rightarrow a^{n-1} \end{cases}$$

The other possibility gives the rules:

$$\text{If } n=2p+1 \quad \begin{cases} (a^{-1})^{p+1} \rightarrow a^p \\ a^{p+1} \rightarrow (a^{-1})^p \end{cases}$$

$$\text{If } n=2p \quad \begin{cases} (a^{-1})^{p} \rightarrow a^p \\ a^{p+1} \rightarrow (a^{-1})^{p-1} \end{cases}$$

In the general case, we mentioned in chapter 2 that the uniform termination problem was undecidable. But we can ask ourselves if the special case of groups is not decidable. The answer is no, for the proof uses a rewriting system associated to a Turing machine, whose operators are unary together with a single constant. This proof obviously applies in the present context of the word rewriting systems.

However, we present now a few proof techniques that will be used in chapter 6. We suppose that the unit element 1 is less than every non-empty word. Frequently, a direct reasoning on the aspect of the rules is necessary.

For example, if an element g from $G \cup G^{-1}$ is such that when it appears in a left member, then the number of its occurrences in this word is greater than in the corresponding right member, that is for all $\lambda \rightarrow \rho$, $n(\lambda,g) \geq n(\rho,g)$ where $n(M,g)$ is the number of occurrences of g in M, then all infinite chains of reduction $w_1 \rightarrow w_2 \rightarrow \cdots \rightarrow w_n \rightarrow \cdots$ are such that there exist two integers i and k with $\forall j > i\ n(w_j,g) = k$. Consequently, every w_j is of the form $v_0^j g v_1^j \cdots v_{k-1}^j g v_k^j$, where $n(v_i^j,g)=0$. Thus a lexicographic ordering on the $k+1$ tuples of words may be used for example.

Finally, when the rules are not length-increasing, the proposition

$$\forall M,M',\ |M|=|M'|\ \lambda_i \text{ left member, } \rho_j \text{ right member, } \rho_j M \not\Rightarrow \lambda_i M'. \qquad \text{(P)}$$

implies the noetherianity: let us suppose that reduction chains such that $\max_i |w_i| < p$ are all finite, then let $w_1 \rightarrow w_2 \rightarrow \cdots \rightarrow w_n \rightarrow \cdots$ be a reduction chain with $\max_i |w_i| = p$ and $(n_j)_j$ the sequence of left member occurrences. We have two cases.

1) $\exists k$ such that $\forall j \geq k\ n_j > 1$,
   Then $\exists c \in G$ with $\forall j \geq k\ w_j = cw'_j$.
   Therefore, the induction hypothesis applied to the w'_j concludes the proof.
2) Else let k be the first index with $n_k = 1$, then
   $w_k = \lambda_m M$ and $w_{k+1} = \rho_m M$.
   Applying the initial proposition, we have $\forall j > k+1\ n_j > 1$ so that we are in case 1.

Thus all reduction chains must halt after a finite number of steps.

3.3 ABELIAN CASE

3.3.1 Semigroups and monoids

The associative-commutative extension of the completion proposed by Peterson and Stickel [Pet81] is now applied to commutative varieties. In non-commutative cases, we examined the normal forms under the variety's canonical system. Now, this is achieved by associative-commutative equivalence of terms.

In the case of semigroups, the associative-commutative equivalence is solved by the structure of *AC*-terms developed in §2.3. A term of a commutative semigroup or monoid on the set $G = \{g_1, \ldots, g_p\}$ is, in accordance with the previously mentioned theory and with the usual notation, the flat structure $g_{i_1} + \cdots + g_{i_j}$, whose *one* equivalent term is
$t = g_{i_1} + (g_{i_2} + (\cdots + g_{i_j}) \cdots)$.
As we did in the non-commutative case, we now choose a new data structure. In the *AC*-unification, an ordering of the flat structure is taken from a total ordering of the

operator domain. Such a total ordering on G gives us the new structure: we represent terms by abelian words $(a_1, \ldots, a_p)$ in $\mathbb{N}^p - \{0\}$, a_j being the occurrence number of g_j in t.

Definition 3.9

Let S be the free commutative semigroup on $G=\{g_1, \ldots, g_p\}$. S is isomorphic to $\mathbb{N}^p - \{0\}$, the set of p-tuples of positive integers with the addition defined by:

$$(a_1, \ldots, a_p)+(b_1, \ldots, b_p) = (a_1+b_1, \ldots, a_p+b_p)$$

the p-tuple $(a_1, \ldots, a_p)$ represents the term $g_1+\ldots+g_p$ with a_j occurrences of g_j. $\mathbb{N}^p - \{0\}$ is partially well ordered componentwise by the usual ordering on the natural numbers:

$$(a_1, \ldots, a_p) \ll (b_1, \ldots, b_p) \ iff \ \forall i=1 \ldots p \ a_i < b_i.$$

$(\mathbb{N}^p - \{0\}, \ll)$ is then a complete lattice.

The ordering $\ll$ is compatible with the addition and noetherian. If $M=(a_1, \ldots, a_p)$ and $N=(b_1, \ldots, b_p)$ then $\max(M,N)$ is the p-tuple $(\max(a_1,b_1),\ldots,\max(a_p,b_p))$ (resp. min). In other words, we choose for the free semigroup on G (resp. monoid) the concrete representation $\mathbb{N}^p - \{0\}$ (resp. $\mathbb{N}^p$). More precisely, this is the simpler case of the structure defined to handle the associative-commutative case.

The embedding rules now replace the β-rules of the previous sections for non-commutative cases. Their variables allow the subterm reductions. For example, with the rules

$$\begin{cases} a+a \rightarrow b \\ a+b \rightarrow a \end{cases}$$

superposition of the second rule on the first one is not immediately possible. We need the extension $a+a+x \rightarrow b+x$ giving the critical pair $(b+b, b)$. The same phenomenon as for associative non-commutative theories appears. We may compute with only *word* rules, variables being *implicit*. Once more, the completion in monoids or semigroups has the usual control structure with the following definitions:

— A term is an abelian word.

— The word M matches N iff $M \ll N$, the matching is $N-M=(b_1-a_1, \ldots, b_p-a_p)$.

— M reduces in N by the rule $\lambda \rightarrow \rho$ iff $\lambda \ll M$, and then $N = (M-\lambda)+\rho$.

- The rules $\lambda \rightarrow \rho$ and $\sigma \rightarrow \tau$ superpose iff $\min(\lambda,\sigma) \neq 0$, and we may consider only one critical pair: $P = (\max(\lambda,\sigma)-\rho, \max(\lambda,\sigma)-\tau)$. The others being confluent when P is introduced as a rule, i.e. they are less general: if (M,N) is a critical pair deduced from $P \gg \max(\lambda,\sigma)$, then $M \rightarrow N$ or $N \rightarrow M$ by the rule created with the max critical pair.

Note that the reduction introduces a partially defined operator -, and that the reflexive-transitive closure of the reduction is closely related to what we have in mind as a division. The condition $\min(\lambda,\sigma)$ is the direct translation of the general condition that we do not superpose on variables. Now, the ordering $\ll$ on a finitely generated semigroup S has nice finite properties: namely, there exists no infinite disordered subset of S. A disordered subset is a subset that contains no pair of comparable elements.

Lemma 3.10 (Dickson)

All the disordered sets of S, a finitely generated commutative semigroup are finite.

Proof. The proof [Dic13] is by induction on the number n of generators. The result is trivial for $n=1$, for then $S=\mathbb{N}-\{0\}$ and $\ll \equiv <$. We assume the lemma true for n-1, then if S is generated by n elements, and D is a non-empty disordered subset of S, we may choose an element $a=(a_1,\dots,a_n)$ in D. Then the set $a+S=\{a+s \mid s \in S\}$ is disjointed from D. Now we split the set $S-(a+S)$ into subsets $A_i^b=\{s \mid s \in S \text{ and } s_i=b\}$, for $i=1,\dots,n$ and $b=0,\dots,a_i-1$. From the induction hypothesis, the sets $D \cap A_i^b$ are finite. As there is a finite number of such sets and we have $S=(a+S) \cup \bigcup_{i,b} A_i^b$, we conclude that D is finite ▪

Applying this result to the set of defining relations of a finitely presented semigroup, we have the following result [Cli67],

Theorem 3.11

Every finitely generated commutative semigroup is finitely presented.

Proof. We can restrict the defining relations to be incomparable. Then, we can apply Thm.2.10 to the completion: suppose we have a noetherian infinite set of rules, applying Dickson's Lemma to the infinite set of left members. We obtain a contradiction with the fact that the left members are irreducible as there must exist two comparable left members, the smaller one being therefore reducible. Thus we can state

[Bal79]:

Proposition 3.12

On a finite presentation, the commutative semigroup completion terminates. On a finitely generated commutative semigroup, defined by an infinite set of equations, this completion computes in an infinite time a finite canonical set of rules.

This result is of course true for the monoids, the only slight difference being that the word (0,...,0) representing the unit element is now allowed.

Thus, we find the well-known fact that the ***uniform*** word problem is solvable in these varieties. This was explicitly noted by Malcev and Emilicev [Mal58, Emi63]. We may claim the same note we did for group completion: the semigroup or monoid commutative completion can be done with rules on non-abelian words, introduced in §3.1, by the adjunction of the rules $g_i g_j \to g_j g_i$ for all $i>j$. But then the termination of the completion process is not insured. For example, if $G=\{a,b,c\}$ and S defined by the single equation $ba=1$, then the completion generates infinitely many rules $ac^n b \to c^n$ for $n \in \mathbb{N}$, with the choice $ba \to 1$, $ca \to ac$ and $bc \to cb$.

The complexity of the uniform word problem, solved by the completion algorithm, has been analyzed by Cardoza, Lipton and Meyer [Car76], and by Mayr and Meyer [May81]. Their main result, based on the fact that commutative semigroups with large presentations can be embedded in commutative ones with small presentations, can be stated as follows:

—There exists an algorithm solving the uniform word problem for commutative semigroups which requires exponential space on all inputs.

—Every algorithm solving the uniform word problem for commutative semigroups requires exponential space on infinitely many inputs.

Here the input is a pair of a presentation and two words in the generators, of the presentation. The question to answer is the equality of the words under the given presentation.

3.3.2 Groups

The canonical system for abelian groups is

$$
\mathrm{G}_C \begin{cases} x+0 & \rightarrow x \\ x+(-x) & \rightarrow 0 \\ x+(-y)+y & \rightarrow x \\ -0 & \rightarrow 0 \\ -(-x) & \rightarrow x \\ -(x+y) & \rightarrow (-x)+(-y) \end{cases}
$$

As an abelian group is also a commutative monoid, the previous structure of abelian words is still used. But now, with the the inverse operator, we work in the free abelian group $\mathbb{Z}^p$, the simplification by substraction being the fact of the rule G_C2. The rule G_C3, embedding of G_C2, and the rule G_C5 are both used via superpositions to control the behaviour of the completion in abelian groups. For example, the rule $a+b \rightarrow c$ superposed on G_C3 with the substitution $\sigma=\{(x,a),(y,b)\}$ gives the critical pair $(a,c-b)$. So that we may restrict the left members to only one generator, and consequently, there is no reason to keep the waiting equations as a list E of abelian word pairs rather than a list of abelian words, the difference between the pair members.

More precisely, let the defining equations above the generators g_j, $j=1,\ldots,n$ be $\sum_{j=1}^{n} a_{i_j} g_j = 0$. Then we take a new criterion for the choice in the creation of a rule $\sum_{j=1}^{n} a_{k_j} g_j \rightarrow 0$. This rule will be such that $|a_{k_1}| = \min_{a_{i_1}\neq 0} |a_{i_1}|$. Now we *change* the control structure of the completion algorithm, by permuting two steps, but this operation does not affect the validity of Thm.2.10; whereas we cleared out the set of waiting equations, we now symmetrize the first rule. A superposition with the rule G_C3 under the substitution

$$
\begin{cases} \sigma(x) & = & a_{k_1} g_1 \\ \sigma(y) & = & \sum_{i\neq 1} a_{k_i} g_i \end{cases}
$$

allows us to generate the rule

$$
\mathrm{R}\colon a_{k_1} g_1 \rightarrow \sum_{i\neq 1} \varepsilon a_{k_i} g_i
$$

where ε is the opposite sign of a_{k_1}. Note that there exist many possible choices for the substitution σ, as it is the case in the associative-commutative unification, but here we are interested in the *good* one. If $a_{k_1}<0$, a new superposition on G_C5 with $\sigma(x)=a_{k_1} g_1$ gives a rule whose left hand side coefficient is now positive.

Then a superposition with G_C3 immediately followed by a superposition with G_C5 eliminates the inverse of g_1 (the details of the substitutions are left to the reader):

$$R': -g_1 \rightarrow (a_{k_1}-1)g_1 + \sum_{i\neq 1} a_{k_i} g_i .$$

The set of rules **R** and **R'**, together with the ***extended*** rule of **R** when the coefficient is greater than one (cf. Prop.2.29), is symmetrized, i.e. all its critical pairs with G_C are solved (observe, however, the amount of work performed by associative-commutative completion in order to solve the critical pair of R' on $-(-x)\rightarrow x$!). Of course, these rules are automatically produced while creating **R**, as is the usual extended rule for associative-commutative completion. Moreover, in the present case of abelian groups, the two rules associated to **R** are only used by the reduction, not by the superposition. So that they are entirely implicit or coded in the reduction procedure. The second major improvement in the control structure is now to reduce *all* the waiting equations by the system $G_C \cup \{R, R'\}$. This rewriting yields new equations with coefficients a'_{i_j} such that $|a'_{k_j}| < a'_{k_1}$. In fact these reductions perform an integer division with the number of reductions equal to the quotient and the final coefficient equal to the remainder. The reductions generalize the usual integer division (or division on an euclidean ring) to any finitely presented algebra. For example, a *division* was introduced in the commutative semigroup by the partially defined operator -, but it was a division on tuples, here we may restrict our attention to only one generator. After this reduction step, two situations may occur:

1) One of the a'_{j_1}, $j \neq k$ is non-null, the selection of a new rule R_1 is initiated, and the process goes on.
2) All the a'_{j_1}, $j \neq k$ are null, that is, the generator g_1 has disappeared from the waiting equations and the process goes on with the following generator.

Obviously, this algorithm halts, ***without superposition between the generated rules***, with a final rewriting system of the form:

$$T \begin{cases} b_1 g_{i_1} & \rightarrow \sum_{j>i_1}^{n} c_j^1 g_j \\ etc & \cdots \\ b_m g_{i_m} & \rightarrow \sum_{j>i_m}^{n} c_j^m g_j \end{cases}$$

such that $0 \leq |c_i^j| < b_i$, $i_j < i_{j+1}$, as the rules are inter-reduced.

Here, the symmetrization transforms each presentation into a standard one. We shall see that this is also true for the non-abelian groups in chapter 5. Some observations may be done on the previous system. First when the generator g_k does not appear in a left member, $\forall n \in \mathbf{Z}$, $n.g_k$ is in canonical form, i.e. the subgroup generated by g_k is a free abelian group of rank one (a copy of $\mathbf{Z}$), direct component of G, the original group. Because left members are reduced to a single generator, the reductions act independently from the neighboring generators. Second, when all the c_j^k are null, the generator in the corresponding left member generates a group isomorph with $\mathbf{Z}/ b_k \mathbf{Z}$.

Now, as this completion algorithm naturally reduces to a symmetrization one, its cost also restricts, at constant cost for the search of minimal element, to the price of the reductions. So we seek for a limitation of these reductions. Moreover, our selection of the next rule to create was arbitrarily done. Thus we will remove these two points if we choose the new rule to be the one with the minimal non-null absolute value among all the waiting equations. This smaller coefficient settles the left member, so that in one pass we will do a lot of work by reductions. Thus the selection of the next rule may now give a left member generator distinct from the previous one. The second case where the generator disappears from the waiting equations may also be improved, let this rule be

$$\mathrm{R}_p:\ a_j g_j \rightarrow \sum_{i \neq j} b_i g_i$$

and let us divide all the b_i by a_j: $b_i = c_i a_j + d_i$. We have the equation:

$$a_j(g_j - \sum_{i \neq j} c_i g_i) = \sum_{i \neq j} d_i g_i$$

And we may look at this equation as introducing a new generator g'_j:

$$\begin{cases} g_j \rightarrow g'_j - \sum_{i \neq j} c_i g_i \\ a_j g'_j \rightarrow \sum_{i \neq j} d_i g_i \end{cases}$$

And this modification of the given presentation does not affect the waiting equations as they do not contain g_js. This way of introducing new generators has been studied by J. Pedersen in an attempt to solve uniformly the word problem for some varieties [Ped84]. Now, the old generator being eliminated, we keep its dependency as a rule, and the last rule is introduced with the waiting equations as all the d_is are smaller than the last computed minimum. This computation contains two special cases:

1) $a_j = \pm 1$, the generator g_j was dependent from the other ones. So far it is extra work to introduce a new generator, we just keep trace of this dependency with this rule.

2) $d_i = 0$ for $i \neq j$, then the new generator defines a finite cyclic group $\mathbb{Z}/a_j\mathbb{Z}$. The equation $a_j g'_j = 0$ is not introduced as a new equation, for it is completely independent now from the remaining generators, i.e. it defines a finite cyclic group, direct component of G.

Thus we have the following abelian group completion algorithm (let us recall that $R(S)$ denotes the set of irreducible forms of elements in the set S under the set of rules R):

Abelian Group Completion Algorithm

Input A finite presentation **G** consisting of a set G *of* n generators g_i
and a set E of abelian words.

$L := E$; The set of waiting equations.

$r := \phi$; The current symmetrized rule.

$R := \phi$; The set of rules.

$p := 0$; Index of the generator with minimum coefficient.

$k := n$; Number of generators.

$G := \{g_1, \ldots, g_n\}$; The set of generators.

Loop {

If $L = \phi$ **Then** *Success with result* (G,R)

Else { *Let* $W = (a_1, \ldots, a_n)$ *be the abelian word from* L *and* p *the index such that* $|a_p|$ *is minimal and non-null in* L;

$r := \{|a_p| g_p \rightarrow \sum_{i \neq p} \varepsilon a_i g_i,$ (Symmetrization)

$-g_p \rightarrow (a_p - 1) g_p + \sum_{i \neq p} a_i g_i \mid \varepsilon \textit{ sign of } a_p\};$

If *there exists a word in* $L - \{W\}$ *with non-zero* p*-th element*

Then $L := r(L - \{W\})$ **%m%** Interreduction.

Else { **If** $|a_p| = 1$ **Then** { $R := r(R) \cup \{g_p \rightarrow \sum_{i \neq p} \varepsilon a_i g_i\};$

$G := G - \{g_p\}; L := r(L - \{W\})$ } **%g%** Elim. of g_p.

Else { **If** $a_i = 0$ *for all* $i \neq p$

Then $R := r(R) \cup \{|a_p| g_p \rightarrow 0, -g_p \rightarrow (|a_p| - 1) g_p\}$ **%g%**

Else { $k := k + 1;$ (Intro. of a new generator).

$r := \{g_p \rightarrow \sum_{i \neq p} -c_i g_i\};$

$W' := (d_1, \ldots, d_{p-1}, a_p, d_{p+1}, \ldots, d_n)$;
where $a_i = c_i a_p + d_i$ *for* $i \neq p$;
If $d_i = 0$ *for* $i \neq p$
Then { $R := r(R) \cup r \cup \{a_p g_k \to 0\}$; $L := L - \{W\}$ } **%g%**
Else { $R := r(R) \cup \{g_p \to g_k - \sum_{i \neq p} c_i g_i\}$;
$L := [L - \{W\}] \cup \{W'\}$ } **%m%**
$G := (G - \{g_p\}) \cup \{g_k\}$ }}}}} ■

The correction of the algorithm as usually follows from the general correction Thm.2.10.

Theorem 3.13

Let G *be a finite abelian group presentation. The abelian group completion halts on input* G. *Let* $|G_1|$ *(resp.* $|G_2|$*) be the set of generators, either belonging to* G *or introduced by the completion, that do not appear (resp. appear with a coefficient* >1*) in final left members. Then, the abelian group presented by* **G** *is the direct product of* $|G_1|$ *infinite cyclic groups and* $|G_2|$ *finite groups whose cardinals are given by the rewriting system R, wich reduces a word given in the primitive generators into this cyclic decomposition of* **G**.

Proof. Proofs of cyclic decompositions can be found in standard textbooks of algebras. Now the algorithm halts necessarily as each loop step necessarily executes one of the labelled instructions, and **%m%** means that the minimum strictly decreases while **%g%** means that the cardinal of G strictly decreases. In the final canonical rewriting system, we have three cases for every generator g, either initial or introduced by the algorithm:

1) It does not appear in left members, it generates an infinite cyclic group as $n.g$ is in canonical form, $\forall\, n \in \mathbb{N}$. Let G_1 be the set of all these generators.
2) It appears in a left member (such a rule is unique), with a coefficient equal to one. Then it was linearly dependent on other generators.
3) It appears in a left member with a coefficient $c > 1$. The right member is then null and the generator g generates a finite cyclic groupc $\mathbf{Z}/c\mathbf{Z}$, the set of normal forms for sums of g being $[0, g, \ldots, (c-1).g]$. Let G_2 be the set of those generators ■

Thus, the decomposition properties of abelian groups are found with a completion-like algorithm. From the canonical form point of view, we must point out that the normal form we get is not only the one of the words, but also of the algebras themselves. Other proofs of the abelian group decomposition may be found in classical algebra textbooks. A proof of the decidability of the theory of abelian groups can be found in a paper by Szmielew [Szm54]. In fact, the cyclic decomposition can be refined to obtain a decomposition using powers of primes cyclic groups, using the Chinese lemma (The abelian subgroup $\mathbb{Z}/pq\mathbb{Z}$ is isomorphic to $\mathbb{Z}/p\mathbb{Z}\times\mathbb{Z}/q\mathbb{Z}$ when p and q are relative primes). However, such isomorphisms are not unique in general, so that there is not a canonical way to code them in an algorithm as we constantly do here. Thus we halt our investigation on the cyclic presentation of a given abelian group depending on this precise presentation. The reader who is familiar with abelian groups will have noticed that reductions in E and the introduction of a new generator correspond respectively to the elementary row and column operations on the matrix E. The first step of the analysis corresponds to the triangularization of the matrix E only with row operations. The system T corresponds to:

$$\begin{bmatrix} b_1 & -c^1_{i_2} & -c^1_{i_3} & \cdot & \cdot & \cdot \\ 0 & b_2 & -c^2_{i_3} & \cdot & \cdot & \cdot \\ 0 & 0 & b_3 & \cdot & \cdot & \cdot \\ \cdot & \cdot & \cdot & & & \end{bmatrix} \begin{bmatrix} g_{i_1} \\ g_{i_2} \\ g_{i_3} \\ \cdot \end{bmatrix} = \begin{bmatrix} 0 \\ 0 \\ 0 \\ \cdot \end{bmatrix}$$

The second step computes a diagonalized matrix by allowing column operations corresponding to the introduction of new generators. The interest of the present proof of abelian group decomposition is that the set of rules R keeps trace of what usually appears as some row and column magic on matrices. In fact, the completion halts without superpositions between closed rules.

Let us see an example (note that when a new generator is introduced, it takes the name of the eliminated generator, the user's responses to the system are in bold face):

Al-Zebra

theory : **group**

Commutative version ? **y**

List of data-files: **(foo)**

Al-Zebra: **complete ab**

Representation of ab1

Generators : a b c d e f

Equations :

```
-3a+-2b+8c+6e+3f == 0
-3a+-3b+9c+d+7e+-1f == 0
7a+-1b+-2c+6d+5e+8f == 0
6a+2b+4c+e+-2f == 0
3a+-1b+5c+2d+7e+6f == 0
9a+4b+8c+3d+7e+2f == 0
```

R1 : d ---> 3a+3b+-9c+-7e+f

R2 : e ---> -6a+-2b+-4c+2f

R3 : f ---> -75a+-32b+-20c

R4 : c ---> -387a+3b

R5 : b ---> 23a+b

R6 : 7b ---> 0

R7 : 1498a ---> 0

Canonical Representation of ab1 : Z/7+Z/1498+0*Z

```
Complete Set of Rewriting Rules *ab1
d ---> 3a+3b+-9c+-7e+f
e ---> -6a+-2b+-4c+2f
f ---> -75a+-32b+-20c
c ---> -387a+3b
b ---> 23a+b
7b ---> 0
1498a ---> 0
```

Runtime : 1.6s, GC : 0.4s

Al-Zebra : **quit**

On this example, the three kind of displayed rules appear clearly. Thc first four rules eliminate four generators. Rule 5 renames generator b (thus occurrence of b in left member denotes a new generator, distinct from b's occurrence in the right member). The others exhibit finite cyclic groups. In abelian groups, the set of rules computes a change of basis.

To conclude this chapter, let us say that the commutative semigroup completion was studied by Ballantyne and Lankford [Bal79], (where they also study unification in commutative semigroups), independently by J.M. Hullot and G. Huet (unpublished manuscript, May 80). D.A. Smith [Smi66] published a detailed algorithm for abelian group decomposition (which inspired the above presentation). Smith's algorithm has been further analyzed by Bradley [Bra71]. The abelian group completion uses euclidean integer division whose exact complexity is yet unknown.

4 Rings, modules and algebras

In this chapter, we examine completion of varieties involving polynomials. These are *two-level* varieties: first, they are defined with two binary operations, a multiplicative one and an additive one; second, there is a set of *coefficients*, the scalars. We first examine rings, then succinctly modules and algebras. The last two structures must be considered as generalizations of abelian groups, which are themselves **Z**-modules, and of rings, which are **Z**-algebras. The techniques of investigation remain those of Chapter 3. We first consider the symmetrization, or restricted completion between a presentation and the canonical system of the variety. Then we analyze the behaviour of critical pair computations.

4.1 RINGS

Let us recall that a ring is an abelian group, together with a multiplicative operation so that a ring possesses a semigroup structure. The multiplication is distributive with respect to the addition. Thus the set F of operator symbols has three elements: addition, unary minus and multiplication.

First of all, we make the assumption that the rings are unitary ones, that is, multiplication has a unit element 1, 0 being the addition neutral element. This assumption is a slight one: the same as the difference between semigroups and monoids. For our purpose, it just implies that we allow in our data structures the empty one. The addition is commutative, not necessarily the multiplication.

4.1.1 Non-abelian rings

The canonical system for non-commutative rings is [Hul80b]:

$$
R_U \left\{
\begin{array}{llll}
x+0 & \to x & & \\
(-x)+x & \to 0 & x+(-y)+y & \to x \\
1.x & \to x & (x.y).z & \to x.(y.z) \\
x.1 & \to x & x.(y+z) & \to (x.y)+(x.z) \\
-0 & \to 0 & (x+y).z & \to (x.z)+(y.z) \\
-(-x) & \to x & -(x+y) & \to (-x)+(-y) \\
x.0 & \to 0 & x.(-y) & \to -(x.y) \\
0.x & \to 0 & (-x).y & \to -(x.y)
\end{array}
\right.
$$

The rules split into two sets: right ones will be used by the symmetrization, while left ones only define the new data structure. The $\mathbf{R}_U$-canonical forms are the usual ones for polynomials. More formally the canonical form of an element P in the ring A with generators G is:

$$P = \sum_i \varepsilon_i a_{i_1}.(\cdots .(a_{i_{n-1}}.a_{i_n})\cdots)$$

where n depends upon i, ε_i denotes an eventual occurrence of the minus operator - applied to the term $S_i = a_{i_1}.(\cdots .(a_{i_{n-1}}.a_{i_n})\cdots)$, and $a_{i_j} \in G \cup V$. The symbol Σ as usual abbreviates a sum. In P, the same term S_i may occur several times, but never occurs twice with one occurrence under the minus operator -, the other one direct component of the sum. The orientation we choose in the canonical system $\mathbf{R}_U$ for the distributive law expresses the classical polynomial linearization. We take a new data structure in bijective correspondence with reduced terms, as we did at the beginning of the previous chapter. This structure is the superposition of two we already know. First, we have an abelian group, such that the first level of the structure will be abelian words. Second, this commutative structure is ***infinitely*** generated by the words from G^*. Lemma 2.20, which solves associative-commutative equivalence, requires an ordering of these words. Since the fundamental operation is superposition, a lexicographic ordering is well-suited. Consequently free ring words may be represented by a decreasing list of non-null words on G^*, with a flag denoting the unary minus. The correspondence is defined on closed terms:

$$\varphi\colon \sum_i \varepsilon_i a_{i_1}.(\cdots .(a_{i_{n-1}}.a_{i_n})\cdots) \longrightarrow \sum_i \varepsilon_i a_{i_1} a_{i_{n-1}} a_{i_n}$$

where $a_{i_j} \in G$. The set $\varphi(T(F \cup G, \phi))$ is noted $\Sigma\langle G\rangle$, its elements will be called polynomials while this terminology is quite confusing in the present context. For brevity, such a polynomial will be noted $\sum_i \varepsilon_i m_i$ where $m_i \in G^*$. The bijection φ acts as an inheritance principle. It defines in an obvious sense on $\Sigma\langle G\rangle$ the addition, multiplication and the opposite of polynomials.

Now, we must specify the behaviour of the associative and commutative equivalence, matching and unification. By the observation following lemma 2.20, we represent the ***abstract*** elements of $\Sigma\langle G\rangle$ by the ***concrete*** representation $\mathbb{Z}\langle G\rangle$ defined by lemma 2.20, when we choose a total ordering $<_G$ of G:

$$\psi \begin{cases} \Sigma\langle G\rangle \quad \rightarrow \quad \mathbb{Z}\langle G\rangle \\ P = \sum_i \varepsilon_i m_i \rightarrow \psi(P) = \sum_{j=1}^{k} \alpha_j m_{i_j} \end{cases}$$

where the word m_{i_j} appears $|\alpha_j|$ times in P. The coefficient α_j is negative when the minus operator - is applied to m_{i_j}, the words m_{i_j} are in increasing order with respect to the *degree* and lexicographic ordering of words (cf. § 3.1.1). What is this degree? We have defined as usual new keywords:

— A word on G is called a monomial; we add the null monomial 0.

— A term is now a polynomial (via $\psi \circ \varphi$).

— The length of a monomial is called its degree, with $|1|=0$, $|0|=-\infty$.

— The degree of a polynomial is the maximum degree of its monomials.

Note that the function ψ is not a bijection, but a surjection, so that in the set $\psi^{-1}(\psi(P))$ we can pick up one element, say $\varepsilon_1 m_{i_1}+\varepsilon_1 m_{i_1}+ \cdots +\varepsilon_k m_{i_k}+\varepsilon_k m_{i_k}$, where ε_l is the sign of α_l and the "term" $\varepsilon_l m_l$ occurs $|\alpha_l|$ times. This function will be called ψ^{-1} by abuse of notation.

We may speak of the heading monomial of P. We have solved the *AC*-equivalence problem What about unification (superposition) and matching (reduction)? On the most general polynomial in $\mathbf{Z}<G>$, these two problems are technically difficult as we must use the diophantine equations of § 2.2. This fact leads to the following choice of symmetrization.

From the addition structure, we get a symmetrization, but still more complex than the abelian group one. However, let us suppose that we have a rule **k**: $\sum_i \alpha_i m_i \rightarrow \sum_j \beta_j n_j$. We superpose this rule on $x+(-y)+y \rightarrow x$. This superposition is done on a term associated to **k** left member via the function $\varphi^{-1}.\psi^{-1}$. On these terms let us take the substitution

$$\sigma(x)=\varepsilon_1 m_1+(\cdots +(\varepsilon_1 m_1+\varepsilon_1 m_1) \cdots) \ |\alpha_1| \text{ times.}$$

$$\sigma(y)=\varepsilon_2 m_2+(\cdots +(\varepsilon_k m_k+\varepsilon_k m_k) \cdots) \text{ the rest of the left member.}$$

then we get a critical pair that may give in $\mathbf{Z}<G>$ the rule:

$$\mathbf{l}:\ \alpha_1 m_{i_1} \rightarrow \sum_{j \neq 1} \alpha_j m_{i_j}$$

In other words, we can *choose* any monomial as left member of the rules. Note that if α_1 was a negative integer, then a superposition with the rule $-(x+y) \rightarrow (-x)+(-y)$ and reductions using rule $-(-x) \rightarrow x$ produce a rule whose left member will have a positive integer coefficient (if this coefficient is -1 a single superposition on $-(-x) \rightarrow x$ is sufficient). And, still with $x+(-y)+y \rightarrow x$, if $\alpha_1 > 1$, the substitution $\{<x, m_{i_1}>, <y, (\alpha_1-1)m_{i_1}>$ creates the rule:

$$\mathbf{m}:\ -m_{i_1} \rightarrow (\alpha_1-1)m_{i_1}+\sum_{j \neq 1}(-\alpha_j)m_{i_j}$$

The symmetrization then halts by producing rules with variables.

Lemma 4.1

The two following sets of rules are symmetrized with respect to $\mathbf{R}_U$ *:*

$$(1)\begin{cases}\mathbf{k}\colon\ m \to \sum_i \alpha_i m_i \\ \mathbf{l}\colon\ mx \to \sum_i \alpha_i m_i x\end{cases} \quad m \neq m_i,\ x \text{ variable.}$$

$$(2)\begin{cases}\mathbf{k}\colon\ \alpha m \to \sum_i \alpha_i m_i \\ \mathbf{l}\colon\ -m \to (\alpha-1)m + \sum_i(-\alpha_i)m_i \\ \mathbf{m}\colon\ \alpha m x \to \sum_i \alpha_i m_i x \\ \mathbf{n}\colon\ \alpha x m \to \sum_i \alpha_i x m_i \\ \mathbf{o}\colon\ \alpha x m y \to \sum_i \alpha_i x m_i y \\ \mathbf{p}\colon\ -mx \to (\alpha-1)mx + \sum_i(-\alpha_i)m_i x \\ \mathbf{q}\colon\ -xm \to (\alpha-1)xm + \sum_i(-\alpha_i)x m_i \\ \mathbf{r}\colon\ -xmy \to (\alpha-1)xmy + \sum_i(-\alpha_i)x m_i y\end{cases} \quad m \neq m_i,\ \alpha>1,\ x,y \text{ variables.}$$

Proof. We first eliminate the case $|m|=1$, which means that a generator was redundant. We reason on first order terms (so that the lemma must be understood with meta-notations). Let $m=a_1.(\cdots(a_{n-1}.a_n)\cdots)$. Then, in case (1), the only superposition for **k** is with the associativity rule by the substitution $\{<x,a_1>,<y,a_2\cdots a_n>\}$. This gives the critical pair $[a_1.(a_2.(\cdots(a_{n-1}.a_n)\cdots).z),(\sum\alpha_i m_i).z]$. The first member reduces by associativity and the second one by right distributivity to give rule **l** of (1). This rule **l** also superposes only with the associativity rule, which gives the pair $(mxy,\sum\alpha_i m_i xy)$, instance of rule **l**.

In case (2), we have seen that rule **l** is created by **k** and $x+(-y)+y \to x$. Other pairs (if $\alpha>2$) between these two rules are resolved: they are created by the substitution $\{<x,\beta m>,<y,\gamma m>\}$ where $\beta+\gamma=\alpha$, $\beta>1$, $\gamma>1$, giving the critical pair $(\beta m,\sum\alpha_i m_i-\gamma m)$. We have:

$$\sum\alpha_i m_i - \gamma m \stackrel{*}{\to} \sum\alpha_i m_i + \gamma\Big[(\alpha-1)m + \sum(-\alpha_i)m_i\Big] \text{ by rule } \mathbf{l},$$

$$=_{AC}\sum(1-\gamma)\alpha_i m_i + ((\alpha-\beta-1)\alpha+\beta)m \text{ by } \gamma(\alpha-1)=(\alpha-\beta-1)\alpha+beta,$$

$\overset{*}{\rightarrow} \sum(1-\gamma+\alpha-\beta-1)\alpha_i m_i+\beta m$ by rule **k**,

$\overset{*}{\rightarrow} \beta m$ by rule $x+(-y)+y \rightarrow x$.

Afterwards, any critical pair between rules **k** and $-(x+y)\rightarrow(-x)+(-y)$ is also confluent. Any substitution unifying these left members creates the terms $\sum(-\alpha_i)m_i$ and

$\alpha(-m)\overset{*}{\rightarrow} \sum\alpha(-\alpha_i)m_i+\alpha(\alpha-1)m$ by rule **l**,

$\overset{*}{\rightarrow} \sum\alpha(-\alpha_i)m_i+\sum(\alpha-1)\alpha_i m_i$ by rule **k**,

$\overset{*}{\rightarrow} \sum(-\alpha_i)m_i$ by rule $x+(-y)+y \rightarrow y$.

With the rule $-(-x)\rightarrow x$, rule **l** gives the term m and

$\sum\alpha_i m_i+(\alpha-1)(-m)\overset{*}{\rightarrow} \sum\alpha_i m_i+\sum(\alpha-1)(-\alpha_i)m_i+(\alpha-1)^2 m$ by rule **l**,

$\overset{*}{\rightarrow} \sum(\alpha-2)(-\alpha_i)m_i+\sum(\alpha-2)\alpha_i m_i+m$ by rule **k**,

$\overset{*}{\rightarrow} m$ by rule $x+(-y)+y \rightarrow x$.

The remaining superpositions create rules with variables:
—**m** (resp. **n**) with right (resp. left) distributivity.
—**o** with right distributivity and rule **n**.
—**p** (resp. **q**) with $(-x).y \rightarrow -(x.y)$ (resp. $x.(-y)\rightarrow-(x.y)$) and rule **l**.
—**r** with $(-x).y \rightarrow -(x.y)$ and rule **q**.
The reader may check that all remaining superpositions create instances of these six rules ▪

These symmetrized rules allow reduction of a monomial such that $\beta m'mm''$, where $\beta>\alpha$ in a polynomial P, and systematic elimination of unary minus applied to reducible monomials. Applying this process to all the waiting equations, we get a symmetrized set of rules: all critical pairs between them and $\mathbf{R}_U$ are resolved. Of course, in ring completion as in the others, we only keep rules of type **k**. They encode all information needed by reduction. Note that our choice of symmetrized rules is not unique, it depends on the choice of a concrete representation of $\mathbb{Z}/\alpha\mathbb{Z}$. We took the set $[0, \ldots, \alpha-1]$. Another possible choice is $[-p+1, \ldots, 0, \ldots, p+1]$ or $[-p, \ldots, 0, \ldots, p]$ according to α's parity. The main advantage of our choice lies in a uniform cancellation by reduction of the minus operator when applied to a reducible monomial.

Now we can look at the problems of matching and unification. Because of our choice for the symmetrization, these problems are restricted to unification of monomials (superposition) and to matching a monomial against a polynomial (reduction). The next lemma details the critical pairs, thus answering the question of unification.

Lemma 4.2
The two rules

k: $\alpha U \rightarrow P$

l: $\beta V \rightarrow Q$, $\alpha,\beta \in \mathbb{N}^*$, $U,V \in G^*$, $P,Q \in \mathbb{Z}<G>$.

superpose iff $\exists A,B,C$ $B \neq 1$, $AB=U$, $BC=V$, *the critical pair is then*

$$((\max(\alpha,\beta)-\alpha)\ ABC+PC\ ,\ (\max(\alpha,\beta)-\beta)\ ABC+AQ).$$

Proof. Say $\alpha>\beta$, then, by lemma 4.1, rule $\beta xV \rightarrow Q$ superposes on rule $\alpha Uy \rightarrow P$ (x,y variables) by the substitution $\{<x,A>,<y,C>\}$. The unified word αABC reduces on PC and $(\alpha-\beta)ABC+AQ$ ∎

Now the monomial αm matches the polynomial $P=\sum_i^k \alpha_i m_i$ iff there exists $j \leq k$ such that $|\alpha_j|>\alpha$ and two monomials (possibly empty) n_1,n_2 such that $m_j=n_1 m n_2$. If we have a rule $\alpha m \rightarrow \sum \beta_i p_i$, then a reduction step adds to P the polynomial $(\alpha_j-\alpha)m_j+n_1(\sum_l \beta_l p_l)n_2$, and then $\mathbf{R}_U$-normalizing this new polynomial.

The description of ring completion is then achieved. The algorithm may not terminate. Note that we choose a precise symmetrization. Without any care in the superposition process, even the symmetrization may not halt. See the following example:

$$\begin{array}{rcl} A+B & \rightarrow & C \\ (A.x)+(B.x) & \rightarrow & C.x \\ (x.A)+(x.B) & \rightarrow & x.C \\ (x.(y.A))+(x.(y.(B)) & \rightarrow & x.(y.C) \\ & \ldots & \end{array}$$

x,y variables.

4.1.2 Abelian Rings

We briefly describe commutative ring completion. The data-structure consists in two superposed abelian word structures. Thus, the completion is deduced from the previous and commutative monoid ones. Let us recall the canonical system of commutative unitary rings.

$$
R_{CU}\left\{\begin{array}{llll}
x+0 \to x & & & \\
(-x)+x \to 0 & x+(-y)+y \to x \\
1.x \to x & x.(y+z) \to (x.y)+(x.z) \\
-0 \to 0 & -(x+y) \to (-x)+(-y) \\
-(-x) \to x & x.(-y) \to -(x.y) \\
x.0 \to 0 &
\end{array}\right.
$$

We have a symmetrization that enables the choice of the left member. Taking a degree and lexicographical ordering, this left member may be, for example, the heading monomial. The six rules with variables in case (2) of lemma 4.1 are now replaced by two rules, one for **k** and one for **l**, as the multipication is commutative:

Lemma 4.3

The two following sets of rules are symmetrized with respect to $\mathbf{R}_{CU}$ *:*

$$
(1)\left\{\begin{array}{lll}
\mathbf{k}: & m \to \sum_i \alpha_i m_i & \\
\mathbf{l}: & mx \to \sum_i \alpha_i m_i x & m \neq m_i,\ x \text{ variable.}
\end{array}\right.
$$

$$
(2)\left\{\begin{array}{lll}
\mathbf{k}: & \alpha m \to \sum_i \alpha_i m_i & \\
\mathbf{l}: & -m \to (\alpha-1)m+\sum_i(-\alpha_i)m_i & \\
\mathbf{m}: & \alpha m x \to \sum_i \alpha_i m_i x & m \neq m_i,\ \alpha>1,\ x \text{ variable.} \\
\mathbf{n}: & -mx \to (\alpha-1)mx+\sum_i(-\alpha_i)m_i x &
\end{array}\right.
$$

Proof. The proof is that of lemma 4.1, with the difference that rule **l** of (1) is an embedded rule in the sense of Prop.2.29 ∎

Then, as for commutative monoids, we compute the critical pairs with respect to the min-max functions on monomials:

Lemma 4.4

The two rules

$$\mathbf{k}:\ \alpha U \to P$$

$$\mathbf{l}:\ \beta V \to Q\ ,\quad \alpha,\beta\in\mathbb{N}^*,\ U,V\in G^*\cdot P,Q\in\mathbb{Z}<G>.$$

superpose iff $\min(U,V)\neq 1$, *the null degree monomial, the critical pair is then*

$$((\max(\alpha,\beta)-\alpha)\max(U,V)+[\max(U,V)\div U].P\ ,$$

$$(\max(\alpha,\beta)-\beta)\max(U,V)+[\max(U,V)\div V].Q).$$

Proof. The same as lemma 4.2 ▪

In this lemma, as for commutative monoids, a new operation ÷ appears, corresponding to a reduction step. This operation is of course associated with the division modulo an ideal; a reduction computes an equivalent term modulo a sub-structure, sub-monoid, normal subgroup or two-sided ideal in the present case. Now, like previous commutative varieties of semigroups and groups, the completion of commutative rings halts:

Theorem 4.5

Given a noetherian reduction ordering, the completion of a finitely presented commutative ring halts in a finite number of steps.

Proof. Yet another consequence of Th.2.10: suppose that R_∞ is infinite, then the set of left members splits into two sets according to the sign of the integer coefficients. One of them is infinite; applying Dickson's lemma to this set gives two rules not inter-reduced, impossible in R_∞ ▪

Note that the theorem also applies to the case where one keeps some negative coefficients in the left members. Thus the uniform word problem for commutative rings is decidable; this has been known for a long time [Hen22, Her26, Hil90, Kön03].

4.1.3 Distributive Lattices

Before the study of structures over a scalar domain, we describe completion in distributive lattices. Let us recall their canonical set.

$$\mathrm{L}_D \begin{cases} x\wedge(x\vee y) & \to\ x \\ x\vee(y\wedge z) & \to\ (x\vee y)\wedge(x\vee z) \\ x\wedge x & \to\ x \\ x\vee x & \to\ x \end{cases}$$

As for rings, normal forms are described by (informal) polynomials. The meet idempotence identity implies that a monomial appears at most once in a normal form and the join one that no generator appears twice in an irreducible monomial. By the absorption rule, no monomial is *included* in another one. Thus an irreducible term of a free distributive lattice on n generators is a subset E of $P(\{1, \ldots, n\})$, the power set of generators, such that for all x, y in E, $x \not\subset y$. We find the well-known fact that a finitely generated distributive lattice is finite.

We can now describe the symmetrization by the usual (surgical) lemma. Actually, this result is interesting in providing a new type of symmetrization.

Lemma 4.6

The two following sets of rules are symmetrized with respect to L_D :

$$(1)\begin{cases} \bigwedge_i m_i \rightarrow \bigwedge_j n_j \\ \bigwedge_i (m_i \vee x) \rightarrow \bigwedge_j (n_j \vee x) \end{cases}$$

$$(2)\begin{cases} m \rightarrow \bigwedge_j n_j \\ n \rightarrow (\bigwedge_j n_j) \wedge n \end{cases} \quad \text{for } all \ n \subset m.$$

Proof. Straighforward (we have omitted the associative-commutative extensions). The left members in (2) are not inter-reduced and the last right member is not necessarily freely reduced. Three exclusive cases may happen when reducing it :

1) $\exists j$ s.t. $n_j \subset n$: therefore n disappears from this left member term by the absorption rule,

2) $\exists j$ s.t. $n \subset n_j$: all such n_j disappear, the freely reduced term is then $(\bigwedge_{k \in K} n_k) \wedge n$ where $K = \{j \mid n \not\subset n_j\}$,

3) otherwise the left member is L_D-irreducible ∎

The superposition of rules is straighforward by max operations (or equivalently by set intersection corresponding here to matching), and is described by lemma 4.4. From both the finiteness of finitely generated free distributive lattices and the inter-irreducibility of rules, the completion halts in a finite number of steps. Thus the completion solves the uniform word problem for distributive lattices, this result is a special case of a theorem due to T. Evans [Eva51a, Eva69]: every finitely presented lattice is hopfian. However, the completion of a finitely presented distributive lattice will be harder than ring completion: first, the absence of minus operator implies the existence of complex left members; second, the symmetrization of a single rule of type (2) produces a number of rules equal to the left member size. More precisely, Bloniarz, Hunt and Rosenkrantz [Blo84] give lower bounds for the equivalence of terms in some finite, distributive and idempotent algebraic structures, from which it follows that the uniform word problem for distributive lattices is NP-hard.

This contrasts with the case of lattices. No *canonical* form is known. But Whitman's algorithm for free lattices [Whi41] solves the word problem. As usual, we may define on the lattice a partial order by $x \leq y$ iff $x \wedge y = x$. This algorithm is a decision procedure for this antisymmetric partial order, therefore solving the word problem (without computing a canonical form, a normal form exists by taking *the* smaller term in an equivalence class [Grä78]). By adding two clauses, Cosmadakis et al [Cos85a, Cos85b] slightly modify Whitman's procedure to solve the uniform word problem for lattices in polynomial time (this uniform decidability was proved in [Eva51a]). We informally describe this algorithm, which contrasts with those already presented, by a case definition of the ordering $\leq$:

Input: a finite presentation (G,E).

1) $M \leq N$, $N \leq M$ for (M,N) in E.

2) If $M \leq N$ and $N \leq O$, then $M \leq O$.

3) $a \leq a$, a in G.

4) $M \wedge N \leq O$ if $M \leq O$ or $N \leq O$.

5) $M \vee N \leq O$ if $M \leq O$ and $N \leq O$.

6) $M \leq N \wedge O$ if $M \leq N$ and $M \leq O$.

7) $M \leq N \vee O$ if $M \leq N$ or $M \leq O$.

For boolean rings, the reader may check that, by the absence of minus operator and absorption rule, the symmetrization reduces to the introduction of usual extended rules and the superposition is described by lemma 4.4. Thus this completion is simpler than the one for distributive lattices and rings. From the celebrated paper [Coo71], it is NP-hard. This procedure and its extensions to first-order predicate calculus is of importance in theorem-proving. For connections with resolution [Rob65] see e.g. [Hsi82, Fag83a, Pau84, Pau85].

4.2 MODULES AND ALGEBRAS

These algebraic structures are generalizations of abelian groups, $\mathbb{Z}$-modules, and rings, $\mathbb{Z}$-algebras. We restrict our study to the case where the scalar ring is commutative and unitary. Let G_M and G_R be the generators of the module (or G_A for an algebra) and of the scalar ring respectively.

4.2.1 Modules

First of all, we observe, from the canonical system for modules **Mo**, that a normal form of $a.A$, a scalar, A a generator of the module needs that of a. But if B is another module generator, then the normal form of $a.B$ may be $b.B$ (cf. in chapter 2 the canonical system **Mo**). So that an abstract element of the basic ring may have several normal forms: for all generators A, we have surjective maps $a \to a.A$ from the scalar ring into the module, these maps may be non injective. Then, the data structures are those for commutative rings, where scalars now replace integers: the equations are n-tuples of scalar elements: $(a_1, \ldots, a_n)$ stands for the element $\sum_i a_i A_i$ if $G=\{A_1, \ldots, A_n\}$.

As usual, the main part is the search for a good symmetrization. This symmetrization was essentially detailed at the beginning of §3.3.2. Let $\mathbf{k}_0$: $\sum_i a_i A_i \to 0$ be a rule. The a_is are no longer meta-notation abbreviating a_i copies of a monomial; they denote a term. We first isolate a generator, say A_1, as left member. A first superposition of the module rule $y \oplus (\alpha.x) \oplus (\beta.x) \to (\alpha+\beta).x \oplus y$ on $\mathbf{k}_0$, under the substitution $\sigma=\{<\alpha,a_1>, <x,A_1>, <y, \sum_{i\neq 1} a_i A_i>\}$ creates the rule

$$\mathbf{k}_1: (a_1+\beta)A_1 \oplus \sum_{i\neq 1} a_i A_i \to \beta A_1$$

Observe that this is valid only for equations over at least two distinct generators of the module. Otherwise this superposition only gives the extended rule $(a_1+\beta)A_1 \to \beta A_1$, needed to reduce a general term αA_1 where a_1 is a subterm of α.

Now the rule $\mathbf{k}_1$ is superposed on the rule $(-\alpha)+\alpha \to 0$ with $\tau=\{<\alpha,a_1>, <\beta,-a_1>\}$. Observe that this is unification, not just matching. Therefore, the new rule, after **Mo**-reduction is

$$\mathbf{k}_2: -a_1 A_1 \to \sum_{i\neq 1} a_i A_i$$

The notation $-a_1$ abbreviates the **Mo**-normal form of the scalar polynomial coefficient a_1. If this coefficient is reduced to a monomial (n copies of an abelian word on G_R, $n \in \mathbb{N}$), then we are for a while satisfied. Otherwise, the symmetrization goes on by superposing $\mathbf{k}_2$ on $(\alpha.x) \oplus (\beta.x) \to (\alpha+\beta).x$, by $\mu=\{<\alpha,-a_1>, <x,A_1>\}$. And we get

$$\mathbf{k}_3: (-a_1+\beta)A_1 \to \sum_{i\neq 1} a_i A_i + \beta A_1$$

In any case, these types of rules are needed in the general completion algorithm in order to reduce the term $(a_1+\ldots+a_m)A_1$. Now we use the extended rule

$\alpha+(-\gamma)+\gamma \rightarrow \alpha$. If $a_1=\sum_j a_j^1 m_j^1$, where $a_j^1 \in \mathbb{N}$ and m_j^1 abelian word on G_R, then $-a_1=\sum_i -a_i^1 m_i^1$. If we want to select the first monomial, then we use the substitution:

$$\nu \begin{cases} <\alpha, -a_1^1 m_1^1> \\ <\beta, -\sum_{j\neq 1} -a_j^1 m_j^1> \\ <\gamma, \sum_{j\neq 1} -a_j^1 m_j^1> \end{cases}$$

and the next step of the completion may generate the rule

$$\mathbf{k}_4\text{: } -(a_1^1 m_1^1)A_1 \rightarrow \sum_{i\neq 1} a_i A_i + (\sum_{j\neq 1} a_j^1 m_j^1)A_1$$

However, we may wish to suppress this application of the minus operator to the initial coefficient. This is easy but we have two distinct cases: either a_1^1 is equal to 1 or n, with $n>1$, as the corresponding scalars differ by the presence of the + operator. In the first case, with the rule $-(-\alpha) \rightarrow \alpha$, it is straightforward (with the substitution $\lambda=\{<\alpha, m_1^1>\}$, observe that we never superpose on a variable occurrence). For the remaining case, we use the rule $-(\alpha+\beta) \rightarrow (-\alpha)+(-\beta)$, with the substitution $\rho=\{<\alpha, m_1^1>, <\beta, m_1^1+...+m_1^1>\}$. Then the reduction of the critical pair by the previous rule and $-(-\alpha) \rightarrow \alpha$ gives positive left coefficients.

Then, the last step of the symmetrization is, as in the previous section, a matter of taste for the choice of a concrete representation of $\mathbf{Z}/a_1^1\mathbf{Z}$. We take the previous one to reach the final rules

$$\mathbf{k}_5\text{: } a_1^1 m_1^1 A_1 \rightarrow \sum_{i\neq 1} a_i A_i + (\sum_{j\neq 1} a_j^1 m_j^1)A_1$$

$$\mathbf{k'}_5\text{: } -m_1^1 A_1 \rightarrow \sum_{i\neq 1} -a_i A_i + \left(\sum_{j\neq 1} -a_j^1 m_j^1 - (a_1^1 - 1)m_1^1\right)A_1$$

Of course, the second rule exists only if $a_1^1>1$.

Then, the various cases of extended rules including terminal variables are computed in the term completion algorithm, either with the distributive laws as in the ring completion, or with the rules:

$$\begin{cases} (\alpha.x)\oplus(\beta.x)\longrightarrow(\alpha+\beta).x \\ x\oplus(\alpha.x)\longrightarrow(1+\alpha).x \end{cases}$$

In this last case, the extended rules define the new module reduction. The new feature of this reduction is the consequence of the extended rule of $\mathbf{k}_5$:

$$\mathbf{k}_6\text{: } (\alpha+am_1^1)A_1\longrightarrow\sum_{i\neq 1}a_iA_i+(\sum_{j\neq 1}a_j^1m_j^1+\alpha+a'm_1^1)A_1,\ \ \alpha \text{ variable.}$$

They allow the reduction of arbitrary terms where the left member of $\mathbf{k}_5$ occurs: $(bm+\sum_i b_im_i)A_1$ where $|b|>a$ in $\mathbb{N}$, and $m\gg m_1^1$ in the free commutative monoid on G_R (cf. Def.3.9). As usual, we keep only one rule $\mathbf{k}_5$, and we code the others in the matching algorithm implementing reductions.

We have proved that we can restrict by symmetrization our left members to be composed of three objects: a positive integer, an abelian word on G_R and an element of G_M. Observe that we have in fact two symmetrizations: the first with respect to the module generators, then for each such generator, a ring symmetrization with respect to the scalar ring generators. Of course pure scalar rules may exist as we also compute in scalar rings. The main point is that this symmetrization yields the following result:

Theorem 4.7

With the previous symmetrization and a noetherian reduction ordering, the completion of a module presentation over a commutative ring always terminates.

Proof. The proof is as for Thm.4.3. The only difference is that we reason about abelian words with positive coefficients $(n,a_1,\ldots,a_m,b_1,\ldots,b_n)$, where $|G_R|=m$ and $|G_M|=n$. Observe that on these words we have $\sum_i b_i = 1$ ∎

The superpositions are straightforward, they are just ring superpositions on rules sharing a common generator in their left members. Note that more can be said about the control structure, as for abelian groups: fix one generator A, create a new rule whose left member is $a.A$ until A disappears from the waiting equations etc... until the last generator. Let us say that more than one rule with g in its left member may exist, and that if the scalar ring has sufficient properties, then the algorithm may be improved. For example, on an Euclidean ring, the abelian group algorithm may be used, the minimum being defined by the division absolute value.

4.2.2 Algebras

Finally, in algebras over a ring, the completion runs as well. In the commutative case, it terminates. The previous theorem applies without the restriction that $\sum_i b_i = 1$. Thus the uniform word problem for commutative polynomials over rings is decidable. As quoted by Mayr and Meyer [May81], this is implicit in [Hen22, Her26, Hil90, Kön03]. The set of rules is usually called Gröbner bases. They appear in [Hir64] with a non-constructive proof of existence, where they are called Standard bases. Observe that Euclidean algorithm and Gauss elimination method are instances of the completion (resp. univariate polynomials and linear polynomials).

In the other case, the symmetrization may run forever without failure as many total reduction orderings exist. Recall that we choose a precise symmetrization, but other less efficient ways exist. The main difference with the modules is that the left members include words from G_A, and the extended rules possess a variable according to the associativity or associativity and commutativity of the algebra multiplication $*$ (resp. superposition with the associative rule $(x*y)*z \rightarrow x*(y*z)$, and extensions as defined in §2.3.2). These two cases are studied in the literature. G.M. Bergman [Ber78] presents the reductions in non-commutative rings (one of his example is displayed in the next section). The reader will find many applications of the diamond lemma in this paper, the critical pairs are called ambiguities. The generalized Newman lemma (lemma 2.14) is implicit in his paper, with a notion of ambiguities resolvable relative to an ordering. This ordering plays the same part as in lemma 2.14. Among various applications, Bergman proves the Poincaré-Birkhoff-Witt theorem by showing that all ambiguities are resolvable (this theorem exhibits a basis of the universal enveloping algebra A of a Lie algebra over a ring R, basis of A as a free module on R). The only significant difference with our rules is that we have scalar coefficients in the left members, allowing completion of all presentations of algebras.

In 1965, B. Buchberger also discovered the completion of commutative multivariate polynomial rings over fields [Buc65], providing many interesting examples of polynomial completion and of orderings on monomials. In the present setting, the completion solves the polynomial ideal word problem, a fundamental decision problem of algebra. Buchberger developed a system BAS of completion of multivariate polynomials with various fields coefficients, based on both the SAC 2 system [Col80] and the ALDES language [Loo76]. Bruno Buchberger also restricts the left members to unitary monomials. We have seen that this restriction is unnecessary. However, when the scalar ring is a field, this restriction can be used to extend the completion, provided that we possess algorithms for the basic operations in the field: the scalar division is used to keep only unitary monomials in left members. The reader will observe that it is a proper extension of the classical Knuth-Bendix algorithm, as field theory is not equational ($\forall x \neq 0,\ xx^{-1}=1$). However, the question for this extension, where we consider the scalar structure as a black box (using the words of B. Buchberger), lies in the axioms that this black box must satisfy. This interesting study is a proper extension of the classical Knuth-Bendix algorithm.

The above study details how the Peterson-Stickel associative-commutative completion and the symmetrization unify these various algorithms. Independently from the author's work, other attempts to show these connexions have been proposed, based on slightly different point of views. Llopis de Trias [Llo83] analyzes ring completion by a small modification of the associative-commutative completion. Winkler [Win84a,pp.112-126] develops an approach where reductions in the scalar ring are considered as "orthogonal" to those in the algebra, an idea suggested by Kandri-Rody and Kapur [Kan84]. This formalizes the idea of considering the scalar reductions as independent of the actual algebra completion. Lankford and Butler [But84] develop an approach based on Hilbert's basis theorem, where the symmetrization is called the "split" of a polynomial. For a discussion of non-commutative ring completion and differential operators, see the work of Galligo [Gal85].

About complexity, Meyer et al [Car75,Car76,May81] quote that their result about the uniform commutative semigroup word problem also applies to commutative rings over $\mathbb{Q}$:

Theorem 4.8

Let P_i, $i=0,\ldots,n$ be polynomials in $\mathbb{Q}[X]$, any algorithm deciding whether or not P_0 belongs to the ideal generated by the P_is, $i=1,\ldots,n$ requires exponential space for infinitely many polynomials P_i.

Thus the problem is exponential space complete. In practice, lower bounds have been given for the maximal degree in a complete set of rules when the number of *indeterminates* (generators of the algebra) equals two or three. Buchberger [Buch79] has given an exact bound for the bivariate case, further extended to the case of trivariate polynomials by Winkler [Win84b]. Some results about complexity can also be found in [Laz83,Giu85].

4.3 SOME EXAMPLES

We begin with an elementary result of cartesian geometry: the diagonal lines of a rhombus intersect at their centre. The following figure introduces the parameters u_i, i=1,2,3, and the indeterminates x_j, j=1, . . . ,4.

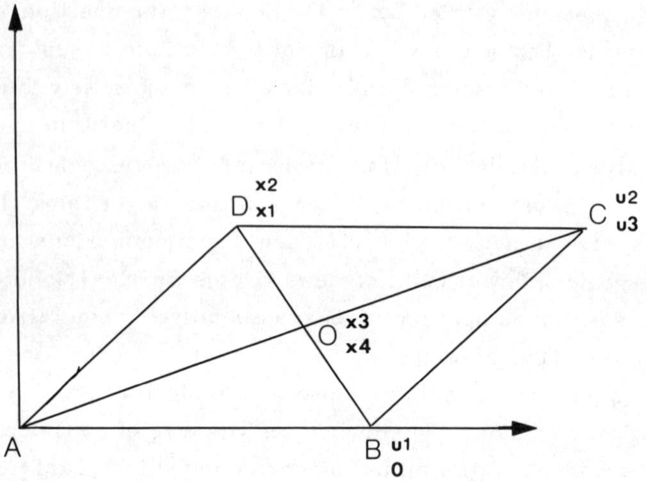

Fig. 4.1

The first two equations give the fourth vertex D of the rhombus. The last ones define the point O as the intersection of the diagonal lines AC and BD.

$$u_1u_3-u_1x_1=0, \quad -u_1x_1+u_2x_1-u_3x_2=0,$$

$$-u_1x_1+u_1x_4+x_1x_3-x_2x_4=0, \quad -u_2x_4+u_3x_3=0.$$

We complete this system (the user's responses are in bold face):

Al-Zebra

theory : **ring**

Commutative version ? **y**

List of data-files: **(examples)**

Al-Zebra: **complete ex1**

Representation of ex1

Generators : u1 u2 u3 x1 x2 x3 x4

Equations :

u1u3 + -u1x1 == 0

-u1x1 + u2x1 + -u3x2 == 0

-u1x1 + u1x4 + x1x3 + -x2x4 == 0

-u2x4 + u3x3 == 0

Mode : **free**

---

u1x1 == u1u3 given

Command ? **y**

R1 : u1x1 ---> u1u3

---

u3x2 == -u1u3 + u2x1 given

Command ? **y**

R2 : u3x2 ---> -u1u3 + u2x1

---

x2x4 == -u1u3 + u1x4 + x1x3 given

Command ? **y**

R3 : x2x4 ---> -u1u3 + u1x4 + x1x3

---

u3x3 == u2x4 given

Command ? **n**

R4 : u2x4 ---> u3x3

2u1u3x3 == u1u2u3 from R4 and R3

Command ? **y**

R5 : 2u1u3x3 ---> u1u2u3

2u1u3x4 == u1u3u3 from R3 and R2

Command ? **y**

R6 : 2u1u3x4 ---> u1u3u3

Complete Set of Rewriting Rules *ex1

u1x1 ---> u1u3

u3x2 ---> -u1u3 + u2x1

x2x4 ---> -u1u3 + u1x4 + x1x3

u2x4 ---> u3x3

2u1u3x3 ---> u1u2u3

2u1u3x4 ---> u1u3u3

Runtime : 2.8s, GC : 0.6s

Al-Zebra: **quit**

The two new equations, computed as critical pairs, can be expressed in the following way:

$$\text{If } u_1u_3 \neq 0, \text{ then } x_3 = \frac{u_2}{2} \text{ and } x_4 = \frac{u_3}{2}.$$

Thus, under the non-degenerate conditions $u_1 \neq 0$ and $u_3 \neq 0$ (when the four points are colinear), the point O is the middle of AC (example inspired from Chon [in Ble84]). This completion is in fact a triangularization as in the abelian group case. The orientation is chosen in order to decrease the degree in the xs of the polynomials. Observe that we explicit only one rule among the four (resp. six) of the symmetrized sets of lemma 4.1 (resp. 4.3. in non-commutative case).

The second example is a computer checking of confluence for a set of rules used by G.M. Bergman [Ber78]. This example proves that in a ring where neither 2 nor 3 are zero divisors, then the four equalities:

$$\begin{cases} a^2 = a \\ b^2 = b \\ c^2 = c \\ (a+b+c)^2 = a+b+c \end{cases}$$

do not imply that $ab=0$. We begin by a completion of the four equations and call R the ring they define:

Al-Zebra

theory : **ring**

Commutative version ? **n**

List of data-files: **(examples)**

Al-Zebra: **complete ex2**

Representation of ex2

Generators : a b c

Equations :

aa == a

bb == b

cc == c

aa + ab + ac + ba + bb + bc + ca + cb + cc == a + b + c

Mode : **free**

aa == a given

Command ? **y**

R1 : aa ---> a

---

bb == b given

Command ? **y**

R2 : bb ---> b

---

cc == c given

Command ? **y**

R3 : cc ---> c

---

cb == -ab + -ac + -ba + -bc + -ca given

Command ? **a** *We ask for another left member.*

ca == -ab + -ac + -ba + -bc + -cb

Command ? **a**

bc == -ab + -ac + -ba + -ca + -cb

Command ? **a**

ba == -ab + -ac + -bc + -ca + -cb

Command ? **y**

R4 : ba ---> -ab + -ac + -bc + -ca + -cb

---

from R1 and R4

cbc == -2ab + -2ac + -bc + -ca + -2cb + -abc + -acb + bca + -cab + -cac

Command ? **a**

cac == -2ab + -2ac + -bc + -ca + -2cb + -abc + -acb + bca + -cab + -cbc

Command ? **a**

cab == -2ab + -2ac + -bc + -ca + -2cb + -abc + -acb + bca + -cac + -cbc

Command ? **a**

bca == 2ab + 2ac + bc + ca + 2cb + abc + acb + cab + cac + cbc

Command ? **y**

R5 : bca ---> 2ab + 2ac + bc + ca + 2cb + abc + acb + cab + cac + cbc

---

Complete Set of Rewriting Rules *ex2

aa ---> a

bb ---> b

cc ---> c

```
ba ---> -ab + -ac + -bc + -ca + -cb
bca ---> 2ab + 2ac + bc + ca + 2cb + abc + acb + cab + cac + cbc
```

Runtime : 6.4s, GC : 0.0s

Al-Zebra : **quit**

The noetherianity of the system follows from two observations [Ber78]:

—The monomials in the right members are either shorter than or of the same length as the one on the left-hand side.

—When a right monomial has the length of its corresponding left member, then it has fewer b s to the left of a, and no more total occurrences of a and b.

Let a,b and c be projections in a vector space; these are idempotents operators. So is their sum if the image subspaces are in direct sum. In finite dimension it can be shown that such operators commute. The complete set above is a simple counter-example in infinite dimension. We observe that the irreducible words are a basis of R as a $\mathbb{Z}$-module, as all left members have no scalar (here scalars=integers). Thus the ring R has no zero divisors, and as ab is irreducible, we have $ab \neq 0$. The answer to the initial question is therefore negative as R provides a counter-example.

5 Dehn algorithm and the symmetrization

This chapter presents a detailed analysis of Dehn algorithm for the word problem in groups [Gre60a] using rewriting techniques. Historically, the word problem was introduced by Dehn in 1911 [Deh11], together with the conjugacy and the isomorphism problems (given two words A,B, does there exist a word X such that $A=XBX^{-1}$, and do two given presentations define the same abstract group?). Independently, Boone [Boo59] and Novikov [Nov55] have proved the existence of a finitely presented group with unsolvable word problem. The original proofs were complex and long (about one hundred pages in [Nov55]). For modern proofs see [Sho67, Lyn77]. Therefore, a natural question was the existence of families of groups with solvable word problem. Many such families are known, e.g. finite, residually finite, abelian, simple groups. Magnus [Mag32] proved that a group with only one defining relation has a solvable word problem; the surface groups are examples. An intensively studied class of groups with solvable word problem is the class of small cancellation groups. These are groups whose length-decreasing rules of a symmetrized presentation imply the limited confluence on the unit element. A key fact is that a syntactic characterization, defined on presentations, exists for such groups. Moreover, the knot groups and the 2-manifold fundamental groups are small cancellation ones. In this chapter, we analyze conditions on the symmetrized rules under which this limited confluence holds. The two basic ideas are a detailed analysis of rules'interdependence, exhibiting diagrams of divergences in reductions and localization in a non-confluent reduction dag. This proof is in fact a case study of the validity conditions of a restricted completion procedure. Another example is provided by boolean algebras, in connection with the resolution principle, as shown by the work of E. Paul [Pau84, Pau85]. The present study exhibits new diagrams on symmetrized sets of relations, rather different from usual ones. The number of such diagrams is not finite, and they are defined with non-length-increasing orderings. But they are more general than usual. For example, the group $G=(A,B,C;ABC=CBA)$ does not satisfy the classical conditions and a non-confluence diagram exists [Gre60]. However, it falls under our conditions if we choose an adequate ordering.

5.1 DEFINITIONS

Let us first recall that we may restrict ourselves to confluence on the unit element (if $W =_G 1$, then $W \xrightarrow{*} 1$) as $W = W' \Leftrightarrow WW'^{-1} = 1$, i.e. restricted confluence over 1 solves the word problem.

From now on, a presentation of a group **G** is a pair (G,S) where G is as usual the set of generators, and S a set of words from the free monoid $(G \cup G^{-1})^*$. These two sets are supposed to be finite, the defining relations of **G** are $W=1$, $W \in S$.

A word W is cyclically reduced iff it is in F-canonical form (F being the canonical set of the variety of groups, cf.§2.4) and if $W = w_1 \cdots w_n$, $w_i \in G \cup G^{-1}$, then w_1 and w_n are not inverses. A relation of **G** is a cyclically reduced word equal to 1 in **G**. A set S of relations is a *symmetrized* set of relations (ssr) if the cyclic permutations of any relation in S and its inverse also belong to S. A *piece* between two distinct relations is a common prefix of these two relations.

Definition 5.1

Let $\mathbf{G}=(G,S)$ *be a group presentation where* S *is a symmetrized set of relations, then*

i) **G** *satisfies the condition* **C'**(n), $n \in \mathbb{N}^*$ *iff all pieces* P *belonging to a relation* W *in* S *are such that* $|P| < \frac{1}{n}|W|$.

ii) **G** *satisfies the condition* **C**(n), $n \in \mathbb{N}^*$ *iff no relation* W *in* S *is the product of fewer than n pieces.*

iii) **G** *satisfies the condition* **T**, *iff for all relations* R_1, R_2, R_3 *not inverse one of the other, then at least one of the products* R_1R_2, R_2R_3, R_3R_1 *is in* F-*canonical form.*

The condition **T** is usually called **T**(4) due to its geometrical interpretation. However, we will drop this classical interpretation. Let us see two examples, first the group presented by $(a,b,c\ ;\ abc, a^3, b^5, c^7)$ is a polyhedral group; it satisfies **C'**(2), but not **C'**(3): a is a piece between the two first relations and $|a| = \frac{|abc|}{3}$. The group $\mathrm{G}=(a,b\,; aba^{-1}b^{-1}ab)$ satisfies **T**. For example, with the triple $(ba^{-1}b^{-1}aba$, $a^{-1}b^{-1}abab$, $b^{-1}a^{-1}b^{-1}a^{-1}ba)$, the product of the first and last relations is in F-canonical form.

The fundamental result of small cancellation theory asserts that, under the **C'**(n) and **T** syntactic hypothesis, Dehn algorithm solves the word problem. This algorithm iteratively replaces in a given word all occurrences of a relation subword whose length is greater than half the whole relation length, together with the F-reduction of course. One immediately observes that this algorithm is a rewriting one:

$$W \xrightarrow[D]{} W' \text{ iff } \exists R\in S,\ U,V,X,Y \text{ s.t. } R=UV, |U|>|V|, W=XUY \text{ and } W'=F(XV^{-1}Y).$$

The relation $\xrightarrow[D]{}$ is generated by the non-inter-reduced rewriting system $D=\{U\rightarrow V \mid \exists R\in S \text{ s.t. } R=UV^{-1}, |U|>|V|\}$.

Theorem 5.2 (Dehn Greendlinger Lyndon)

If a group presentation **G** *satisfies* **C'**(6), *or* **C'**(4) *and* **T**, *then Dehn algorithm solves the word problem for* **G**.

It suffices to prove the confluence on 1 of the relation $\xrightarrow[D]{*}$. We will start in §2 with a detailed analysis of the symmetrization algorithm under the hypothesis of length decreasing rules. As the present study is a detailed analysis of successive reductions, we restrict ourselves to rules that do not increase length.

We also need the usual graph description of a group presentation, known as its Cayley diagram. Let **G** be a group defined on a set G of generators, its Cayley diagram is a directed graph C_G whose vertices are the elements of **G**. The set of edges is the subset of **G**×**G** of all pairs (V,W) such that $W=_{\mathbf{G}}aV$, $a\in G\cup G^{-1}$. Hence, at any vertex W, there are two edges for each generator, one directed towards the vertex, the other directed away. We label the edge (V,W) by the letter a. Consequently, any path corresponds to a word on $(G\cup G^{-1})^{*}$. Let us identify paths and words. A path P is a circuit iff $P=_{\mathbf{G}}1$. Of course, we shall deal only with reduced circuits, having no subpaths like aa^{-1} or $a^{-1}a$. A simple circuit does not cross the same vertex twice. The relations define a set of ***elementary*** circuits. A piece is simply a common subpath between two distinct elementary circuits.

Let us now define S–***diagrams***, S a ssr. The definitions are consistent with [Lyn77]. Let F be a free group and S a ssr on F. Such a diagram D is given by three finite sets (V,E,R) and a function $\varphi: E\rightarrow F-\{1\}$ where V,E,R are subsets of the topological plane. More precisely, we require that:

i) The set V is a set of points, E is a set of open segments, homeomorphic to the open unit interval of the real axis, and R is a set of regions homeomorphic to the open unit disk of the plane. The sets V,E,R are pairwise disjoint. So that the graph (V,E) is planar and the regions are supposed to be oriented as usual for planar graphs.

ii) Let S be a set, $\bar{S}$ is its topological closure and ∂S its frontier $\bar{S}-S$. We have, for all $e\in E$ and $r\in R$, $\bar{e}=e\cup\{a,b\}$ where $a,b\in V$ and $\bar{r}=\bigcup_{i=1}^{n}\bar{e_i}$, with $e_i\in E$. The set $V\cup E\cup R$ is connected and simply connected. An edge $e\in\partial r$ can be oriented according to r's orientation.

iii) The function φ satisfies $\varphi(e)^{-1}=\varphi(e^{-1})$, $e\in E$. It is extended to paths by $\varphi(e_1 \cdots e_n)=\varphi(e_1)\cdots\varphi(e_n)$. For all regions r, $\varphi(r)$ belongs to S.

Thus, the set ∂r can be ordered into an elementary circuit as above, according to r's orientation. Given an S-diagram D, its frontier ∂D is the union of edge closures. An interior region r of D has no boundary edge which is also a boundary edge of D $(\partial r \cap \partial D=\phi)$. A vertex v is interior iff $v \notin \partial D$. In the same way, an interior edge is simply a piece. Therefore, the property **T** states that in an S-diagram no interior vertex has degree three: if **T** does not hold, then there exists a triple of relations $(\alpha S\beta,\beta^{-1}R\gamma,\gamma^{-1}T\alpha^{-1})$, defining the following diagram.

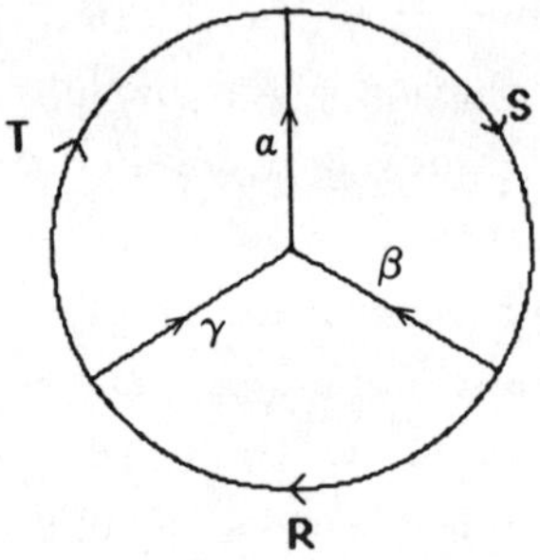

Fig. 5.1

The condition **C**(n) states that an interior vertex v has an exterior degree of at least n, i.e. at least n distinct regions possess v on their boundary. These conditions are localized to either a vertex or a region, we shall define diagrams concerned with sequences of regions.

5.2 THE SYMMETRIZATION

Let us fix some notations. Rule members will be named by greek letters: $\alpha\rightarrow\beta$, $\lambda\rightarrow\rho$, $\mu\rightarrow\nu$, $\tau\rightarrow\sigma\cdots$. Let R be a rewriting system, then the reduction $\xrightarrow[F,R]{}$ was introduced in §4.2, i.e. $\xrightarrow[F]{*}\cdot\xrightarrow[R]{}\cdot\xrightarrow[F]{*}$, while $\xrightarrow[FR]{}$ is the reduction generated by the set of rules $R\cup\{aa^{-1}\rightarrow 1,\ 1a\rightarrow a,\ a1\rightarrow a \mid a\in G\cup G^{-1}\}$.

The symmetrization is just the group completion where the computation of critical pairs other than the normal ones is suppressed (cf. §3.2.1, lemma 3.6). As announced, we restrict ourselves to preorderings on words such that:

i) $\forall A,B\ A>B \Rightarrow \forall U,V\ UAV>UBV$.

ii) $\forall A,B\ |A|>|B| \Rightarrow A>B$.

iii) The preordering > is well-founded.

Let us call them *length orderings*. Such orderings always exist, the simpler one being induced by the word length. However, note that these restrictions do not imply that our ordering is a lexicographic one. On the monoid $\Sigma = \{a,b\}^*$, we define the partial order:

$$1<b<a<bb<ab<ba<aa<bbb<abb<bab<bba<aab<aba<baa<aaa$$

It satisfies all the above properties, but is neither lexicographic nor inverse lexicographic, as $ab<ba$ and $bba<aab$ while $b<a$. It can be extended in a total ordering satisfying all the above properties. There is no restriction to consider the ordering to be total, and we avoid the case of failure. Let us recall the symmetrization algorithm:

Group Symmetrization Algorithm

Input E : A finite set of equations.

$<$: A reduction ordering.

$R_0:=\phi$; The set of rules.

$E_0:=E$; The set of waiting equations.

$RR_0:=\phi$; The set of deleted rules.

$S_0:=\phi$; The set of rules whose canonical pairs are not computed.

$i:=p:=0$; The step counter and the rules counter.

Loop { **If** $RR_i \neq \phi$ **Then** { *choose a pair* $(M_0,N_0)\in RR_i$; $RR_{i+1}:=RR_i - \{(M_0,N_0)\}$ }

If $E_i \neq \phi$ **Then** { *choose a pair* $(M_0,N_0)\in E_i$; $E_{i+1}:=E_i - \{(M_0,N_0)\}$ };

If $RR_i \cup E_i \neq \phi$ **Then** *Create_Rule*$(R_i(M_0),R_i(N_0))$

Else { **If** $S_i=\phi$ **Then** *Success*

Else { *choose a rule* k *in* S_i; $S_{i+1}:=S_i-\{k\}$;

Let E_{i+1} *be the set of canonical pairs of rule* k; $i:=i+1$ }}}

where Create_Rule(M,N) =

If $M=N$ **Then** $i:=i+1$

Else { **If** $M>N$ **Then** { $\lambda:=M$, $\rho:=N$ }

Else $M<N$ **Then** { $\lambda:=N$, $\rho:=M$ };

$p:=p+1$;

$K:=\{k \mid k:\alpha\rightarrow\beta\in R_i$ *and* α *is reducible by* $p:\lambda\rightarrow\rho\}$;

$RR_{i+1}:=\{(\alpha,\beta) \mid k:\alpha\rightarrow\beta\in R_i,\ \&\ k\in K\} \cup RR_i$;

$R_{i+1}:=\{k:\alpha\rightarrow\beta' \mid k:\alpha\rightarrow\beta\in R_i,\ k\notin K,\ \beta'=(R_i\cup\{p:\lambda\rightarrow\rho\})(\beta)\}$

$\cup\{p : \lambda \rightarrow \rho\}$;

$S_{i+1} := S_i \cup \{p\}$;

$i := i+1$ } ∎

The first result about this algorithm is technical. It expresses a continuity between the set R_∞ of stable rules (cf. §2.2.2,Thm2.10).

Lemma 5.3

Let $k: a\lambda b \rightarrow \rho$ *be a rule in* R_∞, $a, b \in G \cup G^{-1}$. *If the presentation* **G** *satisfies* C'(2), *then*

i) $l: a^{-1}\rho \rightarrow \lambda b \in R_\infty$ *or* $a^{-1}\rho \underset{R_\infty}{\longrightarrow} . \overset{*}{\underset{F}{\Rightarrow}} \lambda b$,

ii) $m: \rho b^{-1} \rightarrow a\lambda \in R_\infty$ *or* $\rho b^{-1} \underset{R_\infty}{\longrightarrow} . \overset{*}{\underset{F}{\Rightarrow}} a\lambda$.

Proof. We prove i), as ii) is the symmetrical case. The computation of canonical pairs generates at some iteration the pair $(a^{-1}\rho, \lambda b)$. As k is a stable rule from R_∞ (cf. Thm.2.10), the rule $i: \lambda b \rightarrow a^{-1}\rho$ cannot be generated. This yields only two cases, either the first expressed in i), or the pair has been resolved. In this latter case, there exists $\alpha \neq 1$ and a rule $j: a^{-1}\mu \rightarrow \nu \in R_\infty$ with $a^{-1}\rho = a^{-1}\mu\alpha$, from both facts that ρ is FR_∞-irreducible and, as proved for Thm.2.10, every left member from R_p is R_q-reducible for all $q \geq p$. But our rules are length-decreasing, thus $|a^{-1}\mu| \geq |\nu|$. That is, by assumption C'(2), two of the relations deduced from rules j and k are equal: $a^{-1}\mu\nu^{-1} = a^{-1}\rho b^{-1}\lambda^{-1}$. We have the reductions

$$a^{-1}\rho = a^{-1}\mu\alpha \underset{R_\infty}{\longrightarrow} \lambda b \alpha^{-1}\alpha$$

$$\overset{*}{\underset{F}{\Rightarrow}} \lambda b$$

which proves i) ∎

Proposition 5.4

Given a group presentation **G** *and a length ordering, the symmetrization always terminates. Let* Γ *be the set of rules computed, then if* $k: \lambda \rightarrow \rho \in \Gamma$, *we have :*

- *The words* $\lambda\rho^{-1}$, $\lambda^{-1}\rho$, $\rho\lambda^{-1}$ *and* $\rho^{-1}\lambda$ *are relations.*
- *If* $|\lambda| + |\rho| = 2p+1$ *then* $|\lambda| = p+1, |\rho| = p$.
- *If* $|\lambda| + |\rho| = 2p$ *then* $|\lambda| = p = |\rho|$ *or* $|\lambda| = p+1, |\rho| = p-1$.

If G satisfies C'(2) *then*

- *The set* $S=\{\lambda\rho^{-1}, \rho\lambda^{-1} \mid k: \lambda\rightarrow\rho\in\Gamma\}$ *is the ssr of defining relations.*
- *If* $UV\in S$ *and* $|U|>|V|$ *then* $U \xRightarrow[F,\Gamma]{*} V^{-1}$.
- *A non-confluent critical pair results from a superposition on a piece between two relations.*

Let us remember that the algorithm takes a noetherian ordering as input, so that reductions always terminate.

Proof. We first prove the termination of the procedure by a straightforward adaptation of the first step of Thm.2.10's proof. For each rule $k: \lambda\rightarrow\rho\in R_i$, and all $j\geq i$, there exists another rule $l: \alpha\rightarrow\beta$ in R_j that reduces λ, as for the correctness theorem. Next, the number of words that appear in R_∞ is finite: as no superpositions other than normal ones are allowed, these words are subwords of the symmetrized set of defining relations. Then R_∞ is necessarily finite. By the correctness theorem the symmetrization halts.

Let Γ be the set of rules computed. If $k: \lambda\rightarrow\rho\in\Gamma$, then both left and right hand side are in F-canonical form. If the word $\lambda\rho^{-1}$ was not in F-normal form, then the rule k would be of the type $\lambda'a\rightarrow\rho'a$, $a\in G\cup G^{-1}$. One of its normal pair is $(\lambda', \rho'aa^{-1})$, with F-normal form (λ', ρ'). But all three cases $\lambda'\rightarrow\rho'$, $\rho'\rightarrow\lambda'$ in R, or $\lambda'=\rho'$ contradict either the Γ-irreducibility of rule k or the fact that $\lambda\neq\rho$. For the same reason, the word $\lambda\rho^{-1}$ is cyclically reduced. And the remaining cases are similar.

Set $\lambda=a\lambda'$ in rule k, then the normal pair $(\lambda', a^{-1}\rho)$ is confluent as long as the symmetrization halts. Therefore, λ' being Γ-irreducible, we must have $|a^{-1}\rho|\geq|\lambda'|$. But $|a\lambda'|\geq|\rho|$ by length hypothesis, so that $0\leq|a\lambda'|-|\rho|\leq 2$. Thus, if $|\lambda|+|\rho|=2p$, we have $p\leq|\lambda|\leq p+1$, and if $|\lambda|+|\rho|=2p+1$, we have $p+\frac{1}{2}\leq|\lambda|\leq p+\frac{3}{2}$. This proves the length assertions.

The remaining propositions are consequences of these length inequalities. First, let aR be a word in S, $a\in G\cup G^{-1}$. We know from the first assertion that aR is a relation. From the definition of S, we get $R^{-1}a^{-1}\in S$. Thus we only need to prove $Ra\in S$. Let first aR be of the form $\lambda\rho^{-1}$, then $\lambda=a\lambda'$ and lemma 5.3 implies two cases, either l_1: $a^{-1}\rho\rightarrow\lambda'\in\Gamma$ or there exists $\alpha\neq 1$ such that $\rho=\rho'\alpha$ and l_2: $a^{-1}\rho'\rightarrow\lambda'\alpha^{-1}\in\Gamma$. Involving the definition of the set S on these two rules gives $Ra\in S$. Second, we have $aR=\rho\lambda^{-1}$. We now apply lemma 5.3 twice. First, there exists a rule l_1:$\lambda'b\rightarrow a\rho'\in\Gamma$ where $\lambda'b=\lambda$. It follows that l_2:$a\rho'b^{-1}\rightarrow\lambda'\in\Gamma$. Applying the lemma once more we get the rule we were seeking, l_3:$a^{-1}\lambda'\rightarrow\rho'b^{-1}$. By definition of S, we get in this latter case $Ra\in S$. Thus this set is precisely the symmetrized set of defining relations. Note that we omit in the proof the latter cases of lemma 5.3. The reader can check that this case is exactly the one we studied, modulo subword's labels. Now, if $UV\in S$

and $|U|>|V|$, then

- $UV=\lambda\rho^{-1}$, the length identities imply $|U|\geq|\lambda|$, thus there exists U' such that $\lambda U'=U$ and $U'V=\rho^{-1}$. We have the following derivations:

$$U=\lambda U' \underset{\Gamma}{\Rightarrow} \rho U'=V^{-1}U'^{-1}U' \overset{*}{\underset{F}{\Rightarrow}} V^{-1}$$

- $UV=\rho\lambda^{-1}$ implies $|U|>|\rho|$ and $\exists U'$, $\exists a\in G\cup G^{-1}$ with $\rho aU'=U$ and $aU'V=\lambda^{-1}$. Then, by lemma 5.3, we have the following derivations:

$$U=\rho aU' \overset{*}{\underset{F,\Gamma}{\Rightarrow}} V^{-1}U'^{-1}U' \overset{*}{\underset{F}{\Rightarrow}} V^{-1}$$

Now, to conclude the proof, let (AE,DC) be a critical pair:

$$\begin{cases} k: AB \rightarrow D \\ l: BC \rightarrow E \end{cases} B\neq 1$$

such that the common non-empty subword B is not a piece from S. Two cyclic permutations of the relations associated to rules k and l must be equal. For example:

$$\begin{array}{lll} & BD^{-1}A & = BCE^{-1} \\ \Rightarrow & D^{-1}A & = CE^{-1} \\ \Rightarrow & DD^{-1}AE & = DCE^{-1}E \\ \Rightarrow & AE & =_F DC \end{array}$$

The last equality implies that the critical pair is resolved ▪

Thus, to a rule $\lambda\rightarrow\rho$ we can associate a set of relations, namely $\lambda\rho^{-1}$, its inverse and all cyclic permutations. In this demonstration, we used the identification of two relations associated, modulo cyclic permutations, to two rules sharing a non-empty non-piece subword. This scheme will be used many times later.

Our study of the symmetrization algorithm is now complete. We just add a result on the critical pairs: Prop.5.4 affirms that unresolved critical pairs are deduced by superposition on a piece. If the ssr S satisfies **C'**(4), they are Γ-irreducible.

Proposition 5.5

If S satisies C'(4), *then every critical pair computed from a piece is, perhaps after a F-reduction, in FΓ-normal form.*

Proof We pick up two rules that we superpose on a piece. We first observe that the match includes at most the whole piece by definition of the pieces. We express this by the fact that the piece may be split into three words: the matching subword B, non-empty, a prefix A and a suffix C, both possibly empty, the whole piece being

ABC:

$$\begin{cases} k:\ \alpha AB \rightarrow \beta C^{-1} \\ l:\ BC\gamma \rightarrow A^{-1}\delta \end{cases}$$

The critical pair is $P=(\beta C^{-1}C\gamma,\alpha AA^{-1}\delta)$. It reduces by F in $Q=(\beta\gamma,\alpha\delta)$; words $\beta\gamma$ and $\alpha\delta$ are F-irreducible.

Suppose that the word $\beta\gamma$ is Γ-reducible, the other case being similar. Both β and γ are $F\Gamma$-irreducible. Also, there exists a Γ-rule $m: \lambda\rho \rightarrow \tau$, such that λ is a proper suffix of β and ρ a proper prefix of γ. We have three cases:

1) λ is not a piece between rules m and k, some associated relations are identical:

$$\lambda\rho\tau^{-1}=\lambda C^{-1}B^{-1}A^{-1}\alpha^{-1}\beta_1 \quad \text{where} \quad \beta_1\lambda=\beta \tag{E}$$

Then C is the empty word. Otherwise, ρ and C^{-1} sharing a common prefix by (E), rule l would not be F-reduced at subword $C\gamma$. But the same deduction implies that B or λ is the empty word, which contradicts our assumptions.

2) ρ is not a piece between m and l, the same deduction remains valid:

$$\rho\tau^{-1}\lambda=\rho\gamma_1\delta^{-1}ABC \quad \text{where} \quad \rho\gamma_1=\gamma \tag{E1}$$

Actually, the words λ and C share a common suffix, the $F\Gamma$-irreducibility of βC^{-1} (as right member) implies $C=1$. But k would then be of the type $Ua \rightarrow Va$ and this is impossible by Prop.5.4.

3) λ and ρ must be pieces. We prohibit such cases by supposing that S satisfy **C'**(4). Then $|\lambda\rho|<\frac{1}{2}|\lambda\rho\tau^{-1}|$, but rules do not increase length, and so we get a contradiction. This achieves the proof of Prop.5.5 ∎

The first useful assumption is obvious in view of Prop.5.4 and Prop.5.5:

Hypothesis 1 : The presentation **G** satisfies **C'**(4).

But we keep in mind that, together with the **C'**(2) condition, what we really do not want is the existence in Γ of the two following configurations:

$$\mathbf{L}\begin{cases} \alpha AB \rightarrow \beta\lambda C^{-1} \\ BC\rho\gamma \rightarrow A^{-1}\delta \\ \lambda\rho \rightarrow \tau \end{cases} \qquad \mathbf{R}\begin{cases} \alpha\lambda AB \rightarrow \beta C^{-1} \\ BC\gamma \rightarrow A^{-1}\rho\delta \\ \lambda\rho \rightarrow \tau \end{cases} \qquad \text{Pieces } ABC,\lambda,\rho.$$

They are prohibited by Hypothesis 1.

The symmetrization algorithm generates a normalized or symmetrized presentation Γ with nice syntactic properties. Note that the notion of a symmetrized set of rules or ***normalized presentation*** contains the two notions of Dehn algorithm and of symmetrized set of relations. We already know that the points of divergence in the reductions are critical pairs computed from pieces. The next section is devoted to the analysis of these points. So far they are irreducible. Thus we shall force their reduction by the concatenation of ***contexts***. But let us see first two examples of Γ-rewriting systems: the latter is **C'**(2), the former is not.

The defining relation is $TTA=TTB=1$. The piece TT in fact defines the new relation BA^{-1}. Moreover, the relation TTA is no more of the form $\lambda\rho^{-1}$ or $\rho\lambda^{-1}$ (cf. Prop.5.4), we have lost the fair properties of Γ:

$$\left\{\begin{array}{rcl} A & \rightarrow & B \\ A^{-1} & \rightarrow & B^{-1} \\ TT & \rightarrow & B^{-1} \\ T^{-1}T^{-1} & \rightarrow & B \\ BT & \rightarrow & T^{-1} \\ TB & \rightarrow & T^{-1} \\ B^{-1}T^{-1} & \rightarrow & T \\ T^{-1}B^{-1} & \rightarrow & T \end{array}\right.$$

The following Γ set shows the necessity in the definition of S for both $\lambda\rho^{-1}$ and $\rho\lambda^{-1}$, its defining relation is $ABCD=DCBA$.

$$\left\{\begin{array}{ccc} DCBA & \rightarrow & ABCD \\ BCDA^{-1} & \rightarrow & A^{-1}DCB \\ B^{-1}A^{-1}DC & \rightarrow & CDA^{-1}B^{-1} \\ DA^{-1}B^{-1}C^{-1} & \rightarrow & C^{-1}B^{-1}A^{-1}D \\ D^{-1}C^{-1}B^{-1}A^{-1} & \rightarrow & A^{-1}B^{-1}C^{-1}D^{-1} \\ B^{-1}C^{-1}D^{-1}A & \rightarrow & AD^{-1}C^{-1}B^{-1} \\ BAD^{-1}C^{-1} & \rightarrow & C^{-1}D^{-1}AB \\ D^{-1}ABC & \rightarrow & CBAD^{-1} \end{array}\right.$$

For example, the relation $ABCDA^{-1}B^{-1}C^{-1}D^{-1}$ is not $\lambda\rho^{-1}$ but $\rho\lambda^{-1}$ by the first rule, and $DCBAD^{-1}C^{-1}B^{-1}A^{-1}$ is $\lambda\rho^{-1}$ yet by the first rule, but not $\rho\lambda^{-1}$ (for both Γ sets, termination is trivial).

5.3 CRITICAL PAIR REDUCTION BY CONTEXTS

This section is an analysis of the various cases of critical pair reductions forced by the concatenation of a context to a critical pair. Due to the number of situations the analysis is somewhat tedious. With the previous notations, a piece ABC gives a critical pair $P=(\alpha\delta,\beta\gamma)$ according to the following rules:

$$\begin{cases} k: \alpha AB \to \beta C^{-1} \\ l: BC\gamma \to A^{-1}\delta \end{cases}$$

The members of P are $F\Gamma$-irreducible. We study reductions of this pair P by a left or right context, seeking confluence conditions or irreducibility criteria of the word reduced by the concatenation of the context.

A context is a $F\Gamma$-irreducible word, composed of two parts whose concatenation to a member of the pair P gives a reducible word. A left context is noted μM and a right one $T\tau$; the words M and T, possibly empty, are absorbed by F-reduction, while the words μ and τ are a ***proper*** prefix and a ***proper*** suffix of a Γ-rule left member. They create a Γ-redex after F-cancellation of M or T.

Since the study is symmetrical with respect to $\alpha\delta$ and $\beta\gamma$, we restrict our attention to the word $\beta\gamma$. Reductions start from $\mu M\beta\gamma$ and $\beta\gamma T\tau$, we proceed in three steps:

1) If M (resp. T) reduces the entire word β (resp. γ), then $\mu M\alpha\delta$ (resp. $\alpha\delta T\tau$) has a sequence of reduction closely related to that of $\mu M\beta\gamma$ (resp. $\beta\gamma T\tau$). The members of the critical pair P are confluent in the left (resp. right) context.
2) Study of the reduction beginning at $\mu M\beta\gamma$ when the word β is not absorbed by the F-reduction of M.
3) Study of the reduction of $\beta\gamma T\tau$. Note that this case is not the opposite case of the previous one, as β appears in a right member while γ is subword of the left part of the rule.

5.3.1 Complete absorption of β or γ by F-reduction

Lemma 5.6

If in the reduction by a left (resp. right) context of $\beta\gamma$, β (resp. γ) is absorbed by F-reduction, then, in this context, the critical pair P is confluent.

Proof. The words β and γ appear in the contexts. Therefore we can rewrite them as $\mu M\beta^{-1}$ and $\gamma^{-1}T\tau$. In the left context case, we have

$$\mu M\beta^{-1}\beta\gamma \overset{*}{\underset{F}{\Rightarrow}} \mu M\gamma \tag{E}$$

Therefore, the word $\mu M\beta^{-1}\alpha\delta$ is also reducible by a rule in Γ. We have, by rule $\boldsymbol{k}$, $\beta^{-1}\alpha =_{\mathbf{G}} C^{-1}B^{-1}A^{-1}$, and the second member of this equality being a piece, the hypothesis **C'**(4) (in fact **C'**(2) is enough) together with Prop.5.4 implies the reducibility of $\beta^{-1}\alpha$ by a rule in Γ. Consequently it has a subword that is left member. The corresponding rule must define the same relation up to cyclic permutations

(Assumption **C'**(2)). Thus, we may write (from now on, we freely use without explicit mention the fact that a subword in a rule is $F\Gamma$-irreducible):

$$\exists \lambda_1, \lambda_2, \rho, \beta_1, \alpha_1 \text{ s.t. } \lambda_1^{-1}\beta_1 = \beta,\ \lambda_1 \neq 1,\ \ \lambda_2\alpha_1 = \alpha,\ \lambda_2 \neq 1,$$

$$m: \lambda_1\lambda_2 \rightarrow \rho \in \Gamma \text{, and}$$

$$\lambda_2\rho^{-1}\lambda_1 = \alpha ABC\beta^{-1} = \lambda_2\alpha_1 ABC\beta_1^{-1}\lambda_1$$

We have the following reductions:

$$\beta^{-1}\alpha = \beta_1^{-1}\lambda_1\lambda_2\alpha_1 \xrightarrow[\Gamma]{} \beta_1^{-1}\rho\alpha_1 = \beta_1^{-1}\beta_1 C^{-1}B^{-1}A^{-1}\alpha_1^{-1}\alpha_1$$

$$\xrightarrow[F]{*} C^{-1}B^{-1}A^{-1}$$

Consequently, $\mu M\beta^{-1}\alpha\delta \xrightarrow[F,\Gamma]{*} \mu M\gamma$, which, together with (E), proves the confluence.

The other case is similar. Let us just give the reductions:

$$\beta\gamma\gamma^{-1}T\tau \xrightarrow{*} \beta T\tau$$

and $\alpha\delta\gamma^{-1}T\tau \xrightarrow[F,\Gamma]{*} \alpha ABCT\tau$ by Prop.5.4.

$$\rightarrow \beta C^{-1}CT\tau \text{ by rule k.}$$

$$\xrightarrow[F]{*} \beta T\tau$$

From now on, we suppose that the two words M and T do not completely absorb their neighbour β or γ.

5.3.2 Left Context Reduction

In this section, we may write $\beta = M^{-1}\beta_1$, so that $\mu M\beta\gamma \xrightarrow[F]{*} \mu\beta_1\gamma$. The next reduction is performed by a Γ-rule. Before entering into technical details, let us point out that proofs use the following deduction schemes:

- Identify two relations, and by the consequences of this identity, show that a rule or a context is not in reduced form, or that assumption **C'**(4) is violated.
- Show using the case $U \xrightarrow{*} V$ in Prop.5.4 that a subword in a relation is reducible, enabling once more the confluence of the critical pair P.

The study splits into two cases, according to whether M is empty or not.

5.3.2.1 M≠1

Again, there exist two cases: either the word γ is partially or totally reduced, or it remains unaffected by the reductions. Let $m: \mu\mu_1 \rightarrow \nu\nu_1$ be the rule from Γ that reduces the member of the critical pair. The word ν_1 points out a possible F-reduction.

Lemma 5.7

Given a reduction of $\beta\gamma$ by a left context with F-reduction, then

— When β is cancelled by $F\Gamma$-reductions, the critical pair is confluent in this context.

— Otherwise $\beta\gamma$ reduces to a word φ, F-irreducible, Γ-reducible only with the following configuration :

$$\text{LL}\left\{\begin{array}{lll} k : \alpha AB & \to & M^{-1}\mu_1\nu_1^{-1}\beta_2 C^{-1} \\ l : BC\rho\gamma_1 & \to & A^{-1}\delta \\ m : \mu\mu_1 & \to & \nu_2\lambda\nu_1 \\ n : \lambda\beta_2\rho & \to & \omega \\ \textbf{pieces} : & & ABC,\ \mu_1\nu_1^{-1},\ \beta_2,\ \lambda,\ \rho. \end{array}\right.$$

Proof.

Case 1, the reductions do not cancel a proper subword of β.

We may write $\beta_1=\mu_1\nu_1^{-1}\beta_2$ where $\beta_2\neq 1$. Then

$$\mu M\beta\gamma \xrightarrow[F]{*} \mu\mu_1\nu_1^{-1}\beta_2\gamma$$

$$\xrightarrow[\Gamma]{} \nu\nu_1\nu_1^{-1}\beta_2\gamma$$

$$\xrightarrow[F]{*} \nu\beta_2\gamma=\varphi$$

This last word φ is in F-normal form. First of all, the word μ_1 must be a piece between rules k and m. Otherwise we could write:

$$\mu_1\nu_1^{-1}\nu\mu=\mu_1\nu_1^{-1}\beta_2 C^{-1}B^{-1}A^{-1}\alpha^{-1}M^{-1} \qquad \text{(I)}$$

This identity implies, as $M\neq 1$, that the word μM is F-reducible, which contradicts the fact that a context is $F\Gamma$-irreducible. Thus there exists a piece between k and m, which is $\mu_1\nu_1^{-1}$.

Under what conditions will the word $\varphi=\nu\beta_2\gamma$ be Γ-irreducible? From Prop.5.5, we know that both $\beta_2\gamma$ and $\nu\beta_2$ are $F\Gamma$-irreducible (strictly speaking, in the latter case, this lemma does not apply, but its proof does as the reader may check. One occurrence of β_2 in k's right member rather than in a left one is immaterial here). Thus, a reduction acting on φ must be of the type:

$$n : \lambda\beta_2\rho \to \omega \text{ where } \lambda\neq 1 \text{ and } \rho\neq 1$$

with the following identities:

$$\exists\gamma_1,\nu_2 \text{ with } \rho\gamma_1=\gamma \text{ and } \nu_2\lambda=\nu$$

Thus we get the configuration **LL** in which the already known pieces are ABC and $\mu_1\nu_1^{-1}$. Applying the process of identification of relations three times, we obtain:

– λ is a piece between rules n and m. Otherwise we should have $\nu_1=1$ (rule k is irreducible and $\beta_2\neq 1$ by hyp. of case 1). Then for the same reason $\mu_1=1$, which contradicts the irreducibility of the left context μM.

– β_2 is a piece associated to n and k, otherwise $C=1$ (irreducibility of l and $\rho\neq 1$), furthermore $\nu_1=1$ (irr. of m and $\lambda\neq 1$). Then the two words λ and μ_1 would have a common suffix. This is impossible by application of Prop.5.4 to rule m.

– ρ is a piece between n and l. Otherwise $C=1$ once more (irr. of k). But soon the words B and β_2 have a common suffix, as $B\neq 1$ and $\beta_2\neq 1$ (hyp. of case 1), we have a contradiction (Prop.5.4 and rule k).

In conclusion, if the word φ is Γ-reducible by the rule n, then the three words λ,ρ and β_2 are pieces.

Case 2

In the reductions, the word β is entirely cancelled. This case splits into two sub-cases: μ_1 prefix of β_1, or β_1 prefix of μ_1.

First, if μ_1 is not a piece then we get an identity analogous to (I), implying that the context μM is F-reducible, contradicting the hypothesis. Thus $\mu_1\nu_1^{-1}$, therefore β_1, is a piece, and the assumption **C'**(4) together with case 3 of Prop.5.4 applied to rule k shows that $M\alpha \overset{*}{\underset{F,\Gamma}{\Rightarrow}} \beta_1 C^{-1}B^{-1}A^{-1}$ and the critical pair is confluent by case 3 of Prop.5.4 applied to rule l:

$$\mu M\alpha\delta \overset{*}{\underset{F,\Gamma}{\Rightarrow}} \mu\beta_1 C^{-1}B^{-1}A^{-1}\delta$$

$$\overset{*}{\underset{F,\Gamma}{\Rightarrow}} \mu\beta_1\gamma$$

Second, rule m is now m: $\mu\beta_1\mu_1 \rightarrow \nu\nu_1$ with μ_1 possibly empty. If β_1 is a piece between m and k, then

$$M\alpha \overset{*}{\underset{F,\Gamma}{\Rightarrow}} \beta_1 C^{-1}B^{-1}A^{-1}$$

because $M\alpha ABC\beta_1^{-1}$ is a relation from S and the second member is strictly less longer than the first one. The members of the critical pair are then confluent in the context:

$$\mu M\beta\gamma \overset{*}{\underset{F,\Gamma}{\Rightarrow}} \mu\beta_1\gamma$$

$$\mu M\alpha\delta \overset{*}{\underset{F,\Gamma}{\Rightarrow}} \mu\beta_1 C^{-1}B^{-1}A^{-1}\delta$$

$$\overset{*}{\underset{F,\Gamma}{\Rightarrow}} \mu\beta_1\gamma$$

If β_1 is not a piece, the identity of relations from m and k:

$$\beta_1\mu_1\nu_1^{-1}\nu^{-1}\mu=\beta_1C^{-1}B^{-1}A^{-1}\alpha^{-1}M^{-1}$$

shows that, as $M\neq 1$, the context μM is reducible, contradicting our assumption ■

Let us say that the existence of the contexts **LL** and similar ones is the key to the present proof. There are in fact S-diagrams with length conditions.

5.3.2.2 M=1

The word $\mu\beta\gamma$ is reduced by the rule m: $\mu\mu_1\to\nu\nu_1$. Also the words β and μ_1 have a common prefix.

Lemma 5.8

Given a reduction of $\beta\gamma$ by left context without F-reduction, then

—If the reduction is done by a piece between β and the context, we get a word φ, F-irreducible, Γ-reducible only under the configuration **LL** *(where $M=1$).*

—Otherwise, the critical pair is confluent in this context.

Proof. Again we consider two subcases: whether or not the common prefix between β and μ_1 is a piece.

Case 1, the prefix is a piece.

Necessarily, $\mu_1\nu_1^{-1}$ is a proper prefix of β. Otherwise, hyp. **C'**(4) implies that

$$\alpha \overset{*}{\underset{F,\Gamma}{\Rightarrow}} \beta C^{-1}B^{-1}A^{-1}$$

as β is now a piece. But α is irreducible. We may write $\beta=\mu_1\nu_1^{-1}\beta_1$ where $\beta_1\neq 1$. Then $\mu\beta\gamma\to\nu\beta_1\gamma$, and we are in the first case of lemma 5.7 with the configuration **LL**, where β_1 replaces β_2 and the word M has disappeared.

Case 2, the prefix is not a piece.

We have the identity:

$$\beta C^{-1}B^{-1}A^{-1}\alpha^{-1}=\mu_1\nu_1^{-1}\nu^{-1}\mu \tag{J}$$

Let us suppose that β is a proper prefix of μ_1. There exists a word ε and $\mathbf{a}\in G\cup G^{-1}$ such that $\mu_1=\beta\mathbf{a}\varepsilon$. Then (J) implies the existence of C' such that $C^{-1}=\mathbf{a}C'^{-1}$. As the word $\mu\beta\gamma$ contains the left member of m, these two equalities imply that the rule $\boldsymbol{l}$ is not reduced in its left member $BC\gamma=BC'\mathbf{a}^{-1}\mathbf{a}\gamma_1$ where $\mathbf{a}\gamma_1=\gamma$.

Then μ_1 is a prefix of β: $\beta=\mu_1\beta_1$ and the equation (J) becomes:

$$\mu^{-1}\nu\nu_1\mu_1^{-1}=\alpha ABC\beta_1^{-1}\mu_1^{-1} \tag{J'}$$

Once more, we consider two cases.

If β_1 is the empty word. Then (J') implies

$$\mu^{-1}\nu\nu_1=\alpha ABC \tag{J''}$$

And we have $\nu_1=1$. Effectively, as ν_1 freely cancels a part of γ, rule l would not be F-reduced if both C and ν_1 were non-null. If $c=1$, as $B\neq1$, rule l inter-reduced implies that $\nu_1=1$. And if $C\neq1$, then $\nu_1=1$ necessarily. The rule configuration is now:

$$\begin{cases} l: \ \alpha Ab \rightarrow \beta C^{-1} & \\ k: \ BC\gamma \rightarrow A^{-1}\gamma & \mu^{-1}\nu=\alpha ABC \\ m: \ \mu\beta \quad \rightarrow \nu & \end{cases}$$

In any case we have confluence of the critical pair in context μ:

$$\mu\beta\gamma\rightarrow\nu\gamma \text{ and } \mu\alpha\delta$$

Succinctly, if $|\alpha|>|\mu|$, then $\alpha=\mu^{-1}\alpha'$, $\alpha'\neq1$, then $\nu=\alpha' ABC$ from (J'') and both $\mu\beta\gamma$ and $\mu\alpha\delta$ reduce on $\alpha'\delta$.

If $|\alpha|\leq|\mu|$, then $\mu=\mu'\alpha^{-1}$, μ' possibly empty. Then (J'') gives $\mu'^{-1}\nu=ABC$ and

$$\mu\alpha\delta=\mu'\alpha^{-1}\alpha\delta \xrightarrow[F]{*} \mu'\delta \tag{E1}$$

$$\mu\beta\gamma\rightarrow\nu\gamma \tag{E2}$$

If $|\mu'|\leq|A|$, then $A=\mu'^{-1}A'$, A' possibly empty. We have $\nu=A'BC$ and $\nu\gamma$ reduces to $\mu'\alpha$, with (E1) this gives the desired confluence. Otherwise $|\mu'|>|A|$ implies $\mu'^{-1}=A\mu''^{-1}$, $\mu''\neq1$. Then (J'') gives $\mu''^{-1}\nu=BC$. Then $\mu'\delta$ reduces on $\nu\gamma$ by Prop.5.4 applied to rule l. This with (E2) proves the confluence.

If β_1 is non-empty, then by (J'), β_1^{-1} contains the suffix ν_1. Otherwise, there exist two words $\nu'_1\neq1$ and γ_1 such that

$$\nu_1^{-1}=\beta_1\nu'^{-1}_1, \ \gamma=\nu'^{-1}_1\gamma_1$$

And once more the left member of l would not be reduced at subword $\nu'_1\nu'^{-1}_1$ on the join of C and γ.

Thus, there exists β_2 such that $\beta=\mu_1\beta_1=\mu_1\nu_1^{-1}\beta_2$, where β_2 is possibly empty. Then the equation (J') implies

$$\mu^{-1}\nu=\alpha ABC\beta_2^{-1} \tag{J'''}$$

Now, $\mu\beta\gamma \xrightarrow[F,\Gamma]{*} \nu\beta_2\gamma$. But by definition of the context and rule m, this word is F-irreducible. Thus, in view of (J'''), ν or β_2 is empty.

If ν is empty, then

$\mu\alpha\delta=\beta_2C^{-1}B^{-1}A^{-1}\alpha^{-1}\alpha\delta$ by (J'),

$\xrightarrow[F,\Gamma]{*} \beta_2\gamma$ by Prop.5.4.

and the critical pair is confluent in the context.
Otherwise, β_2=1. Then by (J''') either ν contains the suffix BC so that $\mu\beta\gamma \xrightarrow[F,\Gamma]{*} \nu\gamma$ is further reduced, or it does not and $\nu\gamma$ is irreducible. The reader may check that in both cases, there is confluence. We just sketch the proof:

$|\alpha|>|\mu| \Rightarrow \alpha=\mu^{-1}\alpha'$, confluence on $\alpha'\delta$.

$|\alpha|\leq|\mu| \Rightarrow \mu=\mu'\alpha^{-1}$, μ' possibly empty.

If $|A|\geq|\mu'|$, then confluence on $\mu'\delta$.

If $|A|<|\mu'|$, then confluence on $\nu\gamma$ ∎

5.3.3 Right context reduction

The main difference with respect to the previous study is that reductions of $\beta\gamma$ first reduce γ, subword of ***right*** member, while β is subword of ***left*** member. However, the two analyses are still very closed. The reduction starts with:

$$\beta\gamma T\tau \xrightarrow[F]{*} \beta\gamma_1\tau \quad \text{where} \quad \gamma=\gamma_1 T^{-1}$$

5.3.3.1 T ≠ 1

Lemma 5.9

Given a reduction of $\beta\gamma$ by a right context with F-reduction, then

—If γ is cancelled by $F\Gamma$-reduction, the critical pair is confluent in this context.

—Otherwise $\beta\gamma$ reduces to a word ψ, F-irreducible, Γ-reducible only with a configuration:

$$\text{LR}\begin{cases} k : \alpha AB & \rightarrow & \beta_1\lambda C^{-1} \\ l : BC\gamma_2\sigma_1^{-1}\tau_1 T^{-1} & \rightarrow & A^{-1}\delta \\ m : \tau_1\tau & \rightarrow \textit{above} f \rightarrow & \sigma_1\rho\sigma_2 \\ n : \lambda\gamma_2\rho & & \omega \\ \textbf{pieces} : \gamma_2,\ ABC, & & \sigma_1^{-1}\tau_1,\ \lambda,\ \rho. \end{cases}$$

Proof. As usual we split the proof into two cases.

Case 1

The Γ-reduction by τ does not entirely absorb the word γ. Then $\gamma_1=\gamma_2\sigma_1^{-1}\tau_1$ and m: $\tau_1\tau \rightarrow \sigma_1\sigma$. The word τ_1 is a piece. Otherwise the equality $\sigma_1^{-1}\tau_1\tau\sigma^{-1}=\sigma_1^{-1}\tau_1 T^{-1}\delta^{-1}ABC\gamma_2$ implies that the context $T\tau$ would be reducible.

We have therefore the configuration **LR**. Let us give succinctly the reasons why the three words μ, ν_2, and ν are pieces:

— μ not piece $\Rightarrow C=1$, but, the words μ and B sharing a common suffix, rule k would not be reduced.

— γ_2 not piece $\Rightarrow \sigma_1=1$, but the two words τ_1 and ν soon share a common prefix, rule m would not be reduced.

— ν not piece $\Rightarrow \sigma_1=1$, and τ_1^{-1}, γ_2 have a common suffix, rule l would not be reduced.

Case 2

The Γ-reduction cancels γ. This case is fully symmetrical to the left context case. The rule m becomes $m: \tau_1\gamma_1\tau \rightarrow \sigma_1\sigma$. If γ_1 is not a piece, then $T\tau$ is reducible as proved by the equality

$$\gamma_1\tau\sigma^{-1}\sigma_1^{-1}\tau_1=\gamma_1 T^{-1}\delta^{-1}ABC$$

Thus, γ_1 is a piece between rules m and l, and the critical pair is confluent:

On the one hand $\beta\gamma T\tau \overset{*}{\underset{F}{\Rightarrow}} \beta\gamma_1\tau$.

On the other hand $\alpha\delta T\tau \overset{*}{\underset{F,\Gamma}{\Rightarrow}} \alpha ABC\gamma_1\tau$, because $\delta T \overset{*}{\underset{F,\Gamma}{\Rightarrow}} ABC\gamma_1$ from Prop.5.4 applied to rule l.

We conclude by $\alpha\delta T\tau \overset{*}{\underset{F,\Gamma}{\Rightarrow}} \beta\gamma_1\tau$, by application of rule k.

5.3.3.2 T=1

The word $\beta\gamma\tau$ is reduced by the rule $m: \tau_1\tau \rightarrow \sigma_1\sigma$, hence τ_1 and γ share a common suffix.

Lemma 5.10

Given a reduction of $\beta\gamma$ by right context without F-reduction, then

—If the reduction is done by a piece between γ and the context, then we get a word ψ, $F\Gamma$-irreducible, Γ-reducible only under the configuration **LR** *(where $T=1$).*

—Otherwise, the critical pair is confluent in this context.

Proof.

Case 1

This suffix is a piece between l and m. Then $\tau_1\sigma_1^{-1}$ is a proper γ suffix. Otherwise the left member of l would be the concatenation of two pieces. This is impossible by assumption **C'**(4). We get a configuration, identical to **LR** except in the word T, presently empty. The conclusions still hold.

Case 2

The suffix is not a piece. Then τ_1 is a γ suffix, as in the opposite case, the identity

$$\delta^{-1}ABC\gamma=\tau\sigma^{-1}\sigma_1\sigma^{-1}\tau_1$$

would imply the reducibility of rule k with the common suffix of β and C. We are allowed to write $\gamma=\gamma_1\tau_1$.

1) If γ_1 is the empty word, then the critical pair is confluent (same proof as in case 1 of the previous section).

2) In the other case, γ is of the form $\gamma_2\sigma_1^{-1}\tau_1$ and $\beta\gamma\tau \overset{*}{\underset{F,\Gamma}{\Rightarrow}} \beta\gamma_2\sigma$. The identity deduced from the relations associated to rules l and m gives $\delta^{-1}ABC\gamma_2=\tau\sigma^{-1}$. This implies, by F-irreducibility of $\beta\gamma_2\sigma$ that either γ_2 or σ is empty.

   — If δ is empty then

$$\begin{aligned}\alpha\delta\tau&=\alpha\delta\delta^{-1}ABC\gamma_2\\&\overset{*}{\underset{F}{\Rightarrow}}\alpha ABC\gamma_2\\&\overset{*}{\underset{F,\Gamma}{\Rightarrow}}\beta\gamma_2\end{aligned}$$

   the critical pair is confluent.

   — If γ_2 is empty, then either τ or σ has a sufficiently large subword to initiate the confluence of the critical pair (details left out). This completes the proof of last lemma on reduction by contexts.

To resume the behaviour of a critical pair concatened to a context, we have two possibilities

— Either the confluence of the two members is possible.

— Or we get a term $\varphi=\nu\beta_2\gamma$ (resp. $\psi=\beta\gamma_2\tau$) with β_2 (resp. γ_2)$\neq 1$, F-irreducible, and Γ-reducible only with a configuration **LL** (resp. **LR**) with numerous pieces.

The syntactical form of divergence points in reductions of any word is known. We can now begin the proof of the main theorem 5.2. Of course, we have the configurations **RR** and **RL** corresponding to reductions of the other member of the critical pair $\alpha\delta$.

5.4 CONFLUENCE OF DEHN ALGORITHM

The goal of this chapter is the proof of the classical theorem of small cancellation theory. Let us reformulate it by means of a symmetrized sets of rules.

Theorem 5.11

Let G*=(G,S) be a finite group presentation such that S is a ssr satisfying the conditions* C'(6), *or* C'(4) *and* T. *Then every word in* $(G \cup G^{-1})^*$ *equal in* G *to the unit element has 1 as unique* $F\Gamma$*-irreducible form, where* Γ *is a symmetrized set of rules defined by a length ordering.*

In fact, we shall prove a more general result, based on the non-existence of the configurations **LL**, **LR**, **RL** and **RR** (cf. previous §).

Proof. Let **G** be a finite group presentation. Given a length ordering, the symmetrization procedure computes a finite set of rules Γ described in the previous section.

We still suppose that the ssr S satisfies condition **C'**(4). We shall deal with the reduction $\xrightarrow[F,\Gamma]{*}$ defined by the set of rules

$\Gamma \cup \{a.a^{-1} \to 1, a^{-1}.a \to 1, a.1 \to a, 1.a \to 1 \mid a \in G\}$,

instead of the previous one $\xrightarrow[F\Gamma]{} = \xrightarrow[F]{*}.\xrightarrow[\Gamma]{}.\xrightarrow[F]{*}$. However, the results we proved for this last relation remain true.

Let W be a word such that:

i) $W \xrightarrow[F,\Gamma]{*} 1$,

ii) $\exists W' \neq 1$, F,Γ*–irreducible*, such that $W \xrightarrow[F,\Gamma]{*} W'$.

We define two sets of descendants of W:

$$\Delta(W) = \{ Z \mid W \xrightarrow[F,\Gamma]{*} Z \} \text{ and } \Omega_0(W) = \{ Z \mid W \xrightarrow[F,\Gamma]{*} Z,\ Irred(Z) = \{1\} \}$$

where $Irred(Z)$ is the non-empty set of Z's normal forms under the relation $\xrightarrow[F,\Gamma]{}$. The first step of the proof localizes a point of *strong* divergence in the dag of W's descendants.

Lemma 5.12

There exist two rules k and l from Γ *and two words* U, V *with:*

$$k: \alpha AB \to \beta C^{-1} \qquad l: BC\gamma \to A^{-1}\delta$$

where ABC *is a piece between k and l, both* U *and* V *are* F,Γ*-irreducible,* $Y = U\alpha ABC\gamma V$ *belongs to* $\Delta(W)$, $Y \xrightarrow[F,\Gamma]{} Y_1 = U\beta\gamma V$ *with* $Irred(Y_1) = \{1\}$ *and* $Y \xrightarrow[F,\Gamma]{} Y_2 = U\alpha\delta V$ *with* $1 \notin Irred(Y_2)$.

Proof. We have the inequalities $\{1\} \subset \Omega_0(W) \subset \Delta(W)$ from the hypothesis on the word W (inclusions are strict). As every element in $\Delta(W)$ is less than or equal to W for the well-founded length ordering, we may pick up a maximal element Z_0 in Ω_0. Soon, we have two words Z_1 and Z_2 with the following figure:

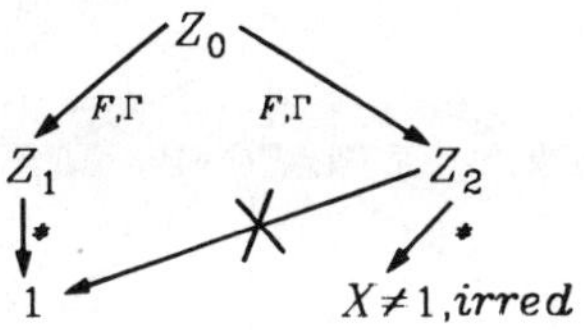

Fig. 5.2

If $1 \in Irred(Z_2)$, then let $\Omega_1(W) = \Omega_0(W) \cap \Delta(Z_2)$. We have the following inclusions:

$$\{1\} \subset \Omega_1(W) \subset \Delta(W) \text{ and } \Omega_1(W) \subset \Omega_0(W)$$

as Z_2 satisfies the hypothesis upon W. By iterating the process, we build a decreasing chain of sets:

$$\Omega_0(W) \supset \Omega_1(W) \supset \cdots \supset \Omega_n(W) \supset \cdots$$

From the noetherianity of the reduction and the finiteness of Γ, we conclude that the set $\Omega_0(W)$ is finite and that the sequence of subsets is stationary from some index such that $\{1\} \not\subset Irred(Z_2)$. Thus, the following set is non-empty:

$$\Theta(W) = \{Z \mid Z \in \Delta(W),\ \exists Z_1, Z_2 \text{ s.t. } Z \xrightarrow[F,\Gamma]{} Z_i,\ \{1\} = Irred(Z_1) \text{ and } 1 \not\in Irred(Z_2)\}.$$

Finally, we choose a minimal word Y in this set $\Theta(W)$ for the reduction ordering.

What are the rules that reduce the word Y on Y_1 and Y_2? They do not belong to F: all the critical pairs of this set are resolved, hence, the pair (Y_1, Y_2) would be confluent. For the same reason, the reductions from Y are done via a non-confluent critical pair of Γ. For in the other case, we would be in case 2 of the Knuth-Bendix algorithm:

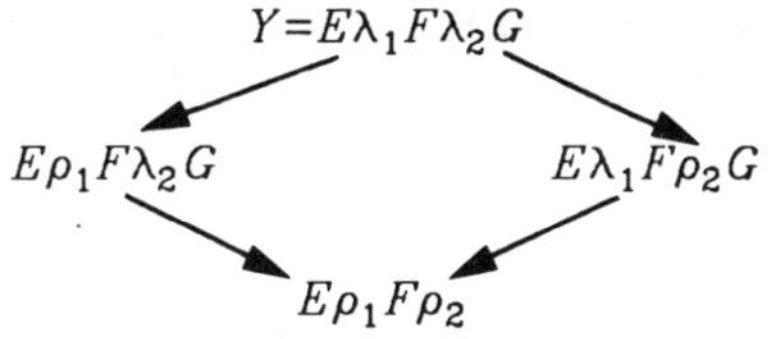

Fig. 5.3

But, from Prop.5.4, we know that an unresolved critical pair is due to a superposition on a piece between two relations. Thus, with the configuration notations of the previous section, we have $Y = U\alpha ABC\gamma V$ with the two rules k and l.

Now, let us suppose that the word U is reducible by a rule i from F or Γ, $U \xrightarrow{i} U'$. The following diagram would contradict the minimality of Y:

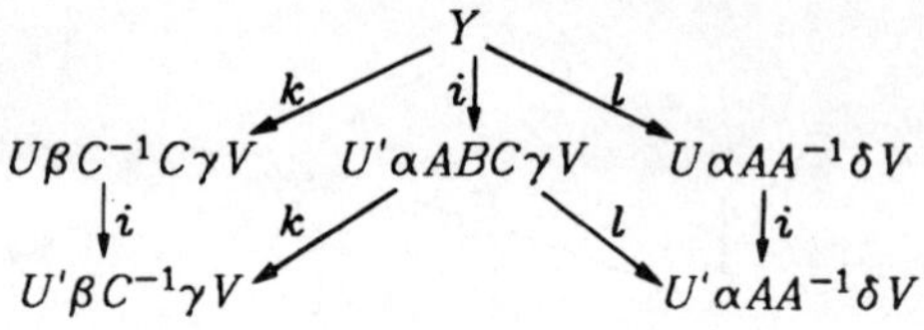

Fig. 5.4

This concludes the proof ∎

Observe that until now, the proof is just a slight modification of the analysis of critical pairs for local confluence of rewriting systems. In the above diagram, the two words $\beta\gamma$ and $\alpha\delta$ behave symmetrically. However, in view of the previous study on symmetrization, let us restrict ourselves to the case where $\beta\gamma$ possesses 1 as unique normal form. After a step of Γ-reduction and perhaps some steps of free cancellation, the word Y reduces on $Y_1 = U\beta\gamma V$ and on $Y_2 = U\alpha\delta V$.

As Y_1 can only reduce on the unit element and U, V, β, γ are F, Γ-irreducible, every reduction must start at either $U\beta$ or γV. Moreover, we can detail the structure of U and V with respect to reductions, as these two words are totally cancelled:

Lemma 5.13

The structure of the words U and V is expressed by the identities :

$$U = \mu_n M_n \cdots \mu_1 M_1$$

$$V = T_1 \tau_1 \cdots T_m \tau_m$$

where the μ_i (resp. τ_j) are prefix (resp. suffix) of left members of rules m_i (resp. n_j) in Γ, non-empty except possibly μ_n (resp. τ_m), such that there exists a reduction path from Y_1 to 1 whose subsequence of Γ-reductions contains the subsequence $(m_i)_{i=1 \cdots n}$ (resp. $(n_j)_{j=1 \cdots m}$) of which each element reduces the corresponding word μ_i (resp. τ_j) in U (resp. V), the words M_i (resp. T_j) expressing possible free cancellations.

We must now analyze precisely this sequence of reductions. In fact, we are going to force these two sequences to be the only possible Γ-reductions.

As said before, the first reduction steps operate on $\mu_1 M_1 \beta\gamma$ or on $\beta\gamma T_1 \tau_1$. None of these two possible reductions gives the confluence of the critical pair as the two reduction dags are totally disconnected. From lemmas 5.6 to 5.10, we deduce the precise nature of the first rewritings: they yield the words φ and ψ. In order to restrict the Γ-reductions, we want these words to be irreducible. Thus our second assumption (after the **C'**(4) condition) asserts the non-existence in Γ of the configurations of the previous section:

$$\mathbf{LL}\left\{\begin{array}{lll} k: \alpha AB & \rightarrow & M_1^{-1}\mu'_1\nu'^{-1}_1\beta_2 C^{-1} \\ l: BC\rho\gamma' & \rightarrow & A^{-1}\delta \\ m: \mu_1\mu'_1 & \rightarrow & \nu_1\lambda\nu'_1 \\ n: \lambda\beta_2\rho & \rightarrow & \omega \\ \textbf{pieces}: \beta_2, & & ABC,\ \mu'_1\nu'^{-1}_1,\ \lambda,\ \rho. \end{array}\right. \qquad \mathbf{LR}\left\{\begin{array}{lll} k: \alpha AB & \rightarrow & \beta'\lambda C^{-1} \\ l: BC\gamma_2\sigma'^{-1}_1\tau'_1 T_1^{-1} & \rightarrow & A^{-1}\delta \\ m: \tau'_1\tau_1 & \rightarrow & \sigma'_1\rho\sigma_1 \\ n: \lambda\gamma_2\rho & \rightarrow & \omega \\ \textbf{pieces}: \gamma_2,\ ABC,\ \lambda, & & \sigma_1^{-1}\tau'_1,\ \rho. \end{array}\right.$$

$$\mathbf{RL}\left\{\begin{array}{lll} k: M_1^{-1}\mu'_1\nu'^{-1}_1\alpha_2 AB & \rightarrow & \beta C^{-1} \\ l: BC\gamma & \rightarrow & A^{-1}\rho\delta \\ m: \mu_1\mu'_1 & \rightarrow & \nu_1\lambda\nu'_1 \\ n: \lambda\alpha_2\rho & \rightarrow & \omega \\ \textbf{pieces}: \alpha_2,\ ABC,\ \lambda, & & \mu'_1\nu'^{-1}_1,\ \rho. \end{array}\right. \qquad \mathbf{RR}\left\{\begin{array}{lll} k: \alpha\lambda AB & \rightarrow & \beta C^{-1} \\ l: BC\gamma & \rightarrow & A^{-1}\delta_2\sigma'^{-1}_1\tau'_1 T_1^{-1} \\ m: \tau'_1\tau_1 & \rightarrow & \sigma'_1\rho\sigma_1 \\ n: \lambda\delta_2\rho & \rightarrow & \omega \\ \textbf{pieces}: \delta_2, & & ABC,\ \sigma'^{-1}_1\tau'_1,\ \lambda,\ \rho. \end{array}\right.$$

Hypothesis 2 : none of the configurations **LL**, **LR**, **RL** and **RR** appears in Γ.

Since the words φ and ψ are irreducible, the next rewriting step reduces either a U suffix or a V prefix. When these reductions occur on the side of $\beta\gamma$ opposed to the first reduction side, we get a word $\nu_1\beta_2\gamma_2\sigma_1$ F-irreducible. Thus, we examine the remaining cases: $\mu_2 M_2\nu_1\beta_2\gamma$ and $\beta\gamma_2\sigma_1 T_2\tau_2$.

The Case $\mu_2 M_2\nu_1\beta_2\gamma$

First, M_2 doesn't freely cancel all ν_1 (M and Ts stand for possible free cancellation). Otherwise, the rule m_1: $\mu_1\mu_1'\rightarrow\nu_1\nu_1'$ where $\mu_1'\nu'^{-1}_1$ is a piece shows that $M_2\mu_1$, with the subword $\nu_1^{-1}\mu_1$ reducible on $\nu_1'\mu_1'^{-1}$ would be reducible. But U is F,Γ-irreducible.

Next, let m_2: $\mu_2\mu_2'\rightarrow\nu_2\nu_2'$, then we have the identity

$$\nu_1 = M_2^{-1}\mu_2'\nu_2'^{-1}\beta_3 \text{ where } \beta_3\neq 1 \text{ and } \mu_2'\nu_2'^{-1} \text{ is a piece.}$$

Suppose a contrario that there exists a prefix P of $\mu_2'\nu_2'^{-1}$ such that $\nu_1=M_2^{-1}P$. Two relations associated to m_1 and m_2 are $P\nu'_1\mu'^{-1}_1\mu_1^{-1}M_2^{-1}$ and $\mu'_2\nu'^{-1}_2\nu_2^{-1}\mu_2$. Then P is not a piece, otherwise the context U would be reducible by $M_2\mu_1 \overset{*}{\underset{F,\Gamma}{\Rightarrow}} P\nu'_1\mu'^{-1}_1$. Thus, the two relations are syntactically equal, this implies first $M_2=1$, second $\mu_2\mu_1$, subword of U, would be F-reducible. Clearly, all these cases are impossible, the existence of the non-empty word β_3 and of the piece $\mu'_2\nu'^{-1}_2$ follows. And we have $\mu_2 M_2\nu_1\beta_2\gamma \overset{*}{\underset{F,\Gamma}{\Rightarrow}} \nu_2\beta_3\beta_2\gamma$.

The Case $\beta\gamma_2\sigma_1 T_2\tau_2$

With the previous reasoning, T_2 does not freely cancel all the word σ_1. And if $n_2 : \tau'_2\tau_2 \rightarrow \sigma'_2\sigma_2$ then $\sigma_1 = \gamma_3\sigma'^{-1}_2\tau'_2 T_2^{-1}$. Note that the relations defined by the two rules n_i, $i=1,2$, are at present:

$$T_2^{-1}\tau_1^{-1}\tau'^{-1}_1\sigma'_1 P \text{ (where } P \text{ is a suffix of } \sigma'^{-1}_2\tau'_2) \quad \text{and} \quad \tau_2\sigma_2^{-1}\sigma'^{-1}_2\tau'_2$$

And the subword $\beta\gamma_2$ remains unmodified by this new reduction:

$$\beta\gamma_2\sigma_1 T_2\tau_2 \overset{*}{\underset{F,\Gamma}{\Rightarrow}} \beta\gamma_2\gamma_3\sigma_2$$

Thus in any case, the second step of Γ-reduction does not affect the the β and γs of the previous Γ-reduction.

And this is true for the next reductions. We started with a word W such that $\{1, W'\} \subset Irred(W)$, where $W' \neq 1$, we want to get a contradiction under some assumptions. At this point, we have a word Y_1 with $Irred(Y_1) = \{1\}$, and the description of Y_1's reductions is possible. A contradiction would be the exhibition of $W' \neq 1$, F,Γ-irreducible, such that $Y_1 \overset{*}{\underset{F,\Gamma}{\Rightarrow}} W'$. For this purpose, it is nearly sufficient to prove that the three words $\nu_2\beta_3\beta_2\gamma$, $\nu_1\beta_2\gamma_2\sigma_1$ and $\beta\gamma_2\gamma_3\sigma_2$ are F,Γ-irreducible.

The three cases will be detailed successively. We first remember the configuration that yields the word and summarize the various deductions establishing the existence of prohibited configurations if we suppose the word reducible.

Reduction of $\nu_1\beta_2\gamma_2\sigma_1$

$$\begin{aligned} \alpha AB &\rightarrow M_1^{-1}\mu'_1\nu'^{-1}_1\beta_2 C^{-1} \\ BC\gamma_2\sigma'^{-1}_1\tau'_1 T_1^{-1} &\rightarrow A^{-1}\delta \\ \mu_1\mu'_1 &\rightarrow \nu_1\nu'_1 \\ \tau'_1\tau_1 &\rightarrow \sigma'_1\sigma_1 \end{aligned}$$

Of course each word of $\nu_1\beta_2\gamma_2\sigma_1$ is irreducible. With the reduction of only two adjacent subwords, let $\lambda\rho \rightarrow \omega$ be as usual this rule. We have the following three **R** or **L** prohibited configurations, with the succinct justification that λ and ρ are pieces (n.r. abbreviates not reduced, meaning that a rule either includes a freely reducible subword or possesses a non-empty prefix or suffix common to both its left and right members).

Reduction of $\nu_1\beta_2$

αAB	$\rightarrow$	$M_1^{-1}\mu'_1\nu'^{-1}_1\rho\beta_2 C^{-1}$	$\lambda\rho ... = \lambda\nu'_1\mu'^{-1}_1 ... \Rightarrow$ rule 1 n.r.
$\mu_1\mu'_1$	$\rightarrow$	$\nu_1\lambda\nu'_1$	$\rho...\lambda = \rho...\mu'_1\nu'^{-1}_1 \Rightarrow$ rule 2 n.r.
$\lambda\rho$	$\rightarrow$	ω	

Reduction of $\beta_2\gamma_2$

$$\begin{array}{rcll} \alpha AB & \rightarrow & M_1^{-1}\mu'_1\nu'^{-1}_1\beta_2\lambda C^{-1} & \lambda C^{-1}B^{-1}...=\lambda\rho...\Rightarrow \text{rule 2 n.r.} \\ BC\rho\gamma_2\sigma'^{-1}_1\tau'_1T_1^{-1} & \rightarrow & A^{-1}\delta & \rho...\lambda=\rho...BC\Rightarrow \text{rule 1 n.r.} \\ \lambda\rho & \rightarrow & \omega & \end{array}$$

Reduction of $\gamma_2\sigma_1$

$$\begin{array}{rcll} BC\gamma_2\lambda\sigma'^{-1}_1\tau'_1T_1^{-1} & \rightarrow & A^{-1}\delta & \lambda\rho...=\lambda\sigma'^{-1}_1\tau'_1...\Rightarrow \text{rule 2 n.r.} \\ \tau'_1\tau_1 & \rightarrow & \sigma'_1\rho\sigma & \rho...\lambda=\rho...\tau'^{-1}_1\sigma'_1\Rightarrow \text{rule 1 n.r.} \\ \lambda\rho & \rightarrow & \omega & \end{array}$$

With the reduction of three adjacent words, we have two cases:

Reduction of $\nu_1\beta_2\gamma_2$

$$\begin{array}{rcll} \alpha AB & \rightarrow & M_1^{-1}\mu'_1\nu'^{-1}_1\beta_2C^{-1} & \lambda\beta_2...=\lambda\nu'_1\mu'^{-1}_1...\Rightarrow \text{rule 3 n.r.} \\ BC\rho\gamma_2\sigma'^{-1}_1\tau'_1T_1^{-1} & \rightarrow & A^{-1}\delta & \beta_2\rho...=\beta_2C^{-1}B^{-1}...\Rightarrow \text{rule 2 n.r.} \\ \mu_1\mu'_1 & \rightarrow & \nu_1\lambda\nu'_1 & \rho...\beta_2=\rho...BC\Rightarrow \text{rule 2 n.r.} \\ \lambda\beta_2\rho & \rightarrow & \omega & \end{array}$$

Reduction of $\beta_2\gamma_2\sigma_1$

$$\begin{array}{rcll} \alpha AB & \rightarrow & M_1^{-1}\mu'_1\nu'^{-1}_1\beta_2\lambda C^{-1} & \lambda\gamma_2...=\lambda C^{-1}B^{-1}...\Rightarrow \text{rule 1 n.r.} \\ BC\gamma_2\sigma'^{-1}_1\tau'_1T_1^{-1} & \rightarrow & A^{-1}\delta & \gamma_2\rho...=\gamma_2\sigma'^{-1}_1\tau'_1...\Rightarrow \text{rule 3 n.r.} \\ \tau'_1\tau_1 & \rightarrow & \sigma'_1\rho\sigma & \rho...\gamma_2=\rho...\tau'^{-1}_1\sigma'_1\Rightarrow \text{rule 2 n.r.} \\ \lambda\gamma_2\rho & \rightarrow & \omega & \end{array}$$

With the reduction of the four words, we have only one case:

$$\begin{array}{rcll} \alpha AB & \rightarrow & M_1^{-1}\mu'_1\nu'^{-1}_1\beta_2C^{-1} & \lambda\beta_2...=\lambda\nu'_1\mu'^{-1}_1...\Rightarrow \text{rule 1 n.r.} \\ BC\gamma_2\sigma'^{-1}_1\tau'_1T_1^{-1} & \rightarrow & A^{-1}\delta & \beta_2\gamma_2...=\beta_2C^{-1}B^{-1}...\Rightarrow \text{rule 2 n.r.} \\ \mu_1\mu'_1 & \rightarrow & \nu_1\lambda\nu'_1 & \gamma_2\rho...=\gamma_2\sigma'^{-1}_1\tau'_1...\Rightarrow \text{rule 4 n.r.} \\ \tau'_1\tau_1 & \rightarrow & \sigma'_1\rho\sigma & \rho...\gamma_2=\rho...\tau'^{-1}_1\sigma'_1\Rightarrow \text{rule 2 n.r.} \\ \lambda\beta_2\gamma_2\rho & \rightarrow & \omega & \end{array}$$

The three rules are **L** or **R** configurations, while the four rules ones are respectively **LL** and **LR** configurations. Thus all corresponding reductions cannot exist. But the last configuration is new. We make the assumption that it does not exist and sketch the two remaining cases.

Reduction of $\nu_2\beta_3\beta_2\gamma$

$$\alpha AB \rightarrow M_1^{-1}\mu'_1\nu'^{-1}_1\beta_2C^{-1}$$

$$\begin{array}{lll} BC\gamma & \rightarrow & A^{-1}\delta \\ \mu_1\mu'_1 & \rightarrow & M_2^{-1}\mu'_2\nu'^{-1}_2\beta_3\nu'_1 \\ \mu_2\mu'_2 & \rightarrow & \nu_2\nu'_2 \end{array}$$

Reduction of two adjacent subwords:

Reduction of $\nu_2\beta_3$

$$\begin{array}{llll} \mu_1\mu'_1 & \rightarrow & M_2^{-1}\mu'_2\nu'^{-1}_2\rho\beta_3\nu'_1 & \lambda\rho...=\lambda\nu'_2\mu'^{-1}_2...\Rightarrow\text{rule 1 n.r.} \\ \mu_2\mu'_2 & \rightarrow & \nu_2\lambda\nu'_2 & \rho...\lambda=\rho...\mu'_2\nu'^{-1}_2\Rightarrow\text{rule 2 n.r.} \\ \lambda\rho & \rightarrow & \omega & \end{array}$$

Reduction of $\beta_3\beta_2$

$$\begin{array}{llll} \alpha AB & \rightarrow & M_1^{-1}\mu'_1\nu'^{-1}_1\rho\beta_2C^{-1} & \lambda\rho...=\lambda\nu'_1\mu'^{-1}_1...\Rightarrow\text{rule 1 n.r.} \\ \mu_1\mu'_1 & \rightarrow & M_2^{-1}\mu'_2\nu'^{-1}_2\beta_3\lambda\nu'_1 & \rho...\lambda=\rho...\mu'_1\nu'^{-1}_1\Rightarrow\text{rule 2 n.r.} \\ \lambda\rho & \rightarrow & \omega & \end{array}$$

Reduction of $\beta_2\gamma$

$$\begin{array}{llll} \alpha AB & \rightarrow & M_1^{-1}\mu'_1\nu'^{-1}_1\beta_2\lambda C^{-1} & \lambda\rho...=\lambda C^{-1}B^{-1}...\Rightarrow\text{rule 2 n.r.} \\ BC\rho\gamma & \rightarrow & A^{-1}\delta & \rho...\lambda=\rho...BC\Rightarrow\text{rule 1 n.r.} \\ \lambda\rho & \rightarrow & \omega & \end{array}$$

Reduction of three adjacent words:

Reduction of $\nu_2\beta_3\beta_2$

$$\begin{array}{llll} \alpha AB & \rightarrow & M_1^{-1}\mu'_1\nu'^{-1}_1\rho\beta_2C^{-1} & \lambda\beta_3...=\lambda\nu'_2\mu'^{-1}_2...\Rightarrow\text{rule 2 n.r.} \\ \mu_1\mu'_1 & \rightarrow & M_2^{-1}\mu'_2\nu'^{-1}_2\beta_3\nu'_1 & \beta_3...\lambda=\beta_3...\mu'_2\nu'^{-1}_2\Rightarrow\text{rule 3 n.r.} \\ \mu_2\mu'_2 & \rightarrow & \nu_2\lambda\nu'_2 & \rho...\beta_3=\rho...\mu'_1\nu'^{-1}_1\Rightarrow\text{rule 2 n.r.} \\ \lambda\beta_3\rho & \rightarrow & \omega & \end{array}$$

Reduction of $\beta_3\beta_2\gamma$

$$\begin{array}{llll} \alpha AB & \rightarrow & M_1^{-1}\mu'_1\nu'^{-1}_1\beta_2C^{-1} & \lambda\beta_2...=\lambda\nu'_1\mu'^{-1}_1...\Rightarrow\text{rule 1 n.r.} \\ BC\rho\gamma & \rightarrow & A^{-1}\delta & \beta_2...\lambda=\beta_2...\mu'_1\nu'^{-1}_1\Rightarrow\text{rule 3 n.r.} \\ \mu_1\mu'_1 & \rightarrow & M_2^{-1}\mu'_2\nu'^{-1}_2\beta_3\lambda\nu'_1 & \rho...\beta_2=\rho...BC\Rightarrow\text{rule 1 n.r.} \\ \lambda\beta_2\rho & \rightarrow & \omega & \end{array}$$

Reduction of the four words:

$$\begin{array}{llll} \alpha AB & \rightarrow & M_1^{-1}\mu'_1\nu'^{-1}_1\beta_2C^{-1} & \lambda\beta_3...=\lambda\nu'_2\mu'^{-1}_2...\Rightarrow\text{rule 3 n.r.} \\ BC\rho\gamma & \rightarrow & A^{-1}\delta & \beta_3\beta_2...=\beta_3\nu'_1\mu'^{-1}_1...\Rightarrow\text{rule 1 n.r.} \\ \mu_1\mu'_1 & \rightarrow & M_2^{-1}\mu'_2\nu'^{-1}_2\beta_3\nu'_1 & \beta_2\rho...=\beta_2C^{-1}B^{-1}...\Rightarrow\text{rule 2 n.r.} \\ \mu_2\mu'_2 & \rightarrow & \nu_2\lambda\nu'_2 & \rho...\beta_2=\rho...BC\Rightarrow\text{rule 1 n.r.} \end{array}$$

$$\lambda\beta_3\beta_2\rho \rightarrow \omega$$

As for the previous case, this last configuration is new. And the first four rules one is also new.

Reduction of $\beta\gamma_2\gamma_3\sigma_2$

$$\begin{aligned} \alpha AB &\rightarrow \beta C^{-1} \\ BC\gamma_2\sigma'^{-1}_1\tau'_1T^{-1}_1 &\rightarrow A^{-1}\delta \\ \tau'_1\tau_1 &\rightarrow \sigma'_1\gamma_3\sigma'^{-1}_2\tau'_2T^{-1}_2 \\ \tau'_2\tau_2 &\rightarrow \sigma'_2\sigma_2 \end{aligned}$$

Reduction of two adjacent subwords:

Reduction of $\beta\gamma_2$

$$\begin{aligned} \alpha AB &\rightarrow \beta\lambda C^{-1} && \lambda\rho...=\lambda C^{-1}B^{-1}...\Rightarrow \text{rule 2 n.r.} \\ BC\rho\gamma_2\sigma'^{-1}_1\tau'_1T^{-1}_1 &\rightarrow A^{-1}\delta && \rho...\lambda=\rho...BC\Rightarrow \text{rule 1 n.r.} \\ \lambda\rho &\rightarrow \omega \end{aligned}$$

Reduction of $\gamma_2\gamma_3$

$$\begin{aligned} BC\gamma_2\lambda\sigma'^{-1}_1\tau'_1T^{-1}_1 &\rightarrow A^{-1}\delta && \lambda\rho...=\lambda\sigma'^{-1}_1\tau'_1...\Rightarrow \text{rule 2 n.r.} \\ \tau'_1\tau_1 &\rightarrow \sigma'_1\rho\gamma_3\sigma'^{-1}_2\tau'_2T^{-1}_2 && \rho...\lambda=\rho...\tau'^{-1}_1\sigma'_1\Rightarrow \text{rule 1 n.r.} \\ \lambda\rho &\rightarrow \omega \end{aligned}$$

Reduction of $\gamma_3\sigma_2$

$$\begin{aligned} \tau'_1\tau_1 &\rightarrow \sigma'_1\gamma_3\lambda\sigma'^{-1}_2\tau'_2T^{-1}_2 && \lambda\rho...=\lambda\sigma'^{-1}_2\tau'_2...\Rightarrow \text{rule 2 n.r.} \\ \tau'_2\tau_2 &\rightarrow \sigma'_2\rho\sigma_2 && \rho...\lambda=\rho...\tau'^{-1}_2\sigma'_2\Rightarrow \text{rule 1 n.r.} \\ \lambda\rho &\rightarrow \omega \end{aligned}$$

Reduction of three adjacent subwords:

Reduction of $\beta\gamma_2\gamma_3$

$$\begin{aligned} \alpha AB &\rightarrow \beta\lambda C^{-1} && \lambda\gamma_2...=\lambda C^{-1}B^{-1}...\Rightarrow \text{rule 2 n.r.} \\ BC\gamma_2\sigma'^{-1}_1\tau'_1T^{-1}_1 &\rightarrow A^{-1}\delta && \gamma_2\rho...=\gamma_2\sigma'^{-1}_1\tau'_1...\Rightarrow \text{rule 3 n.r.} \\ \tau'_1\tau_1 &\rightarrow \sigma'_1\rho\gamma_3\sigma'^{-1}_2\tau'_2T^{-1}_2 && \rho...\gamma_2=\rho...\tau'^{-1}_1\sigma'_1\Rightarrow \text{rule 2 n.r.} \\ \lambda\gamma_2\rho &\rightarrow \omega \end{aligned}$$

Reduction of $\gamma_2\gamma_3\sigma_2$

$$\begin{aligned} BC\gamma_2\lambda\sigma'^{-1}_1\tau'_1T^{-1}_1 &\rightarrow A^{-1}\delta && \lambda\gamma_3...=\lambda\sigma'^{-1}_1\tau'_1...\Rightarrow \text{rule 2 n.r.} \\ \tau'_1\tau_1 &\rightarrow \sigma'_1\gamma_3\sigma'^{-1}_2\tau'_2T^{-1}_2 && \gamma_3\rho...=\gamma_3\sigma'^{-1}_2\tau'_2...\Rightarrow \text{rule 3 n.r.} \\ \tau'_2\tau_2 &\rightarrow \sigma'_2\rho\sigma_2 && \rho...\gamma_3=\rho...\tau'^{-1}_2\sigma'_2\Rightarrow \text{rule 2 n.r.} \end{aligned}$$

$$\lambda\gamma_3\rho \rightarrow \omega$$

Reduction of the four words:

$$\alpha AB \rightarrow \beta\lambda C^{-1} \qquad \lambda\gamma_2\ldots=\lambda C^{-1}B^{-1}\ldots\Rightarrow \text{rule 2 n.r.}$$

$$BC\gamma_2\sigma'^{-1}_1\tau'_1T_1^{-1} \rightarrow A^{-1}\delta \qquad \gamma_2\gamma_3\ldots=\gamma_2\sigma'^{-1}_1\tau'_1\ldots\Rightarrow \text{rule 3 n.r.}$$

$$\tau'_1\tau_1 \rightarrow \sigma'_1\gamma_3\sigma'^{-1}_2\tau'_2T_2^{-1} \qquad \gamma_3\rho\ldots=\gamma_3\sigma'^{-1}_2\tau'_2\ldots\Rightarrow \text{rule 4 n.r.}$$

$$\tau'_2\tau_2 \rightarrow \sigma'_2\rho\sigma_2 \qquad \rho\ldots\gamma_3=\rho\ldots\tau'^{-1}_2\sigma'_2\Rightarrow \text{rule 3 n.r.}$$

$$\lambda\gamma_2\gamma_3\rho \rightarrow \omega$$

As the number of new configurations increases, we interpret them geometrically by S-diagrams. All diagrams sharing a common number of rules will fit into the same geometric interpretation. We suppose that paths in diagrams are ordered as words, with length orderings. A graph associated with a configuration of $n+3$ rules $n>0$ will be denoted by C_n, while the subgraph defined by the same rules except the last one will be denoted by O_n. Two C and O diagrams on the same configuration will be said associate. Each relation in the ssr S defines elementary cycles in the Cayley graph C_G. The boundary of regions in diagrams are these elementary cycles. Every partition of cycles in two oriented paths, one clockwise, the other counter-clockwise, is ordered by the given length ordering. We represent the smaller path of a cycle partitioned by a rule with a dashed line. Let us recall that an interior edge, thus belonging to two distinct cycles, represents a piece between two relations. The first diagram is C_0. It codes all **L** and **R** configurations:

$$\mathbf{L}\left\{\begin{array}{l} \alpha AB \rightarrow \beta\lambda C^{-1} \\ BC\rho\gamma \rightarrow A^{-1}\delta \\ \lambda\rho \rightarrow \omega \end{array}\right. \qquad \mathbf{R}\left\{\begin{array}{l} \alpha\lambda AB \rightarrow \beta C^{-1} \\ BC\gamma \rightarrow A^{-1}\rho\delta \\ \lambda\rho \rightarrow \omega \end{array}\right.$$

We have the two following graphs, equal modulo their labelling:

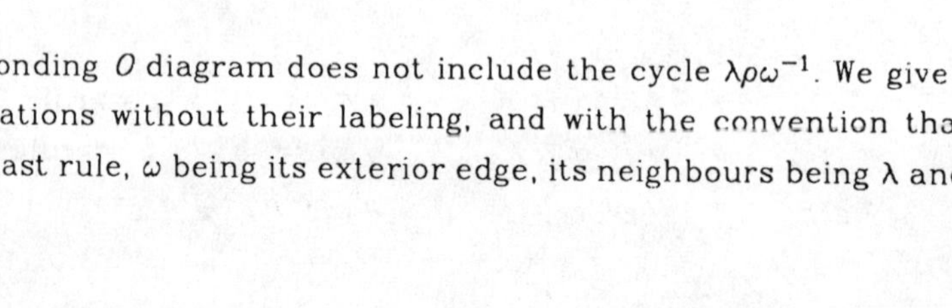

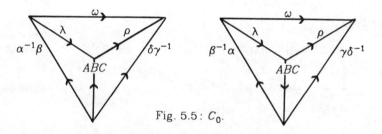

Fig. 5.5: C_0.

The corresponding O diagram does not include the cycle $\lambda\rho\omega^{-1}$. We give the remaining configurations without their labeling, and with the convention that the upper cycle is the last rule, ω being its exterior edge, its neighbours being λ and ρ.

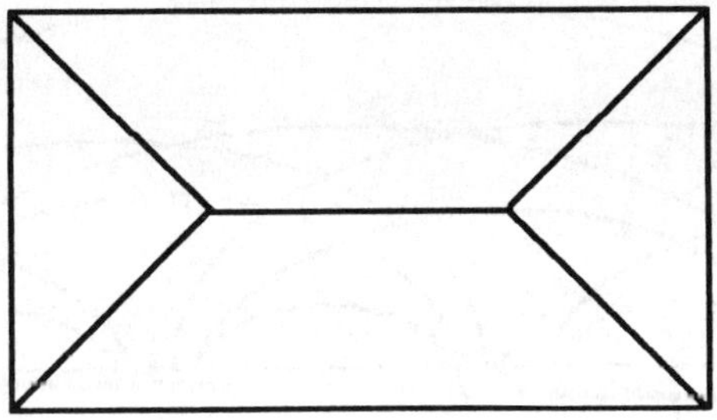

Fig. 5.6: C_1.

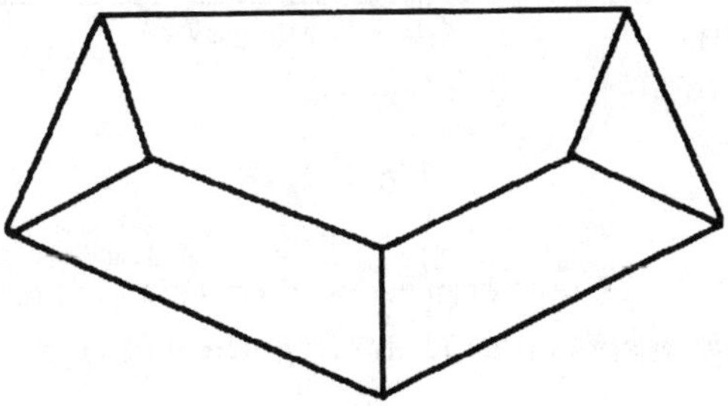

Fig. 5.7: C_2.

Thus, the structure of the following diagrams is obvious. They are composed, for O_n, of n four-gons such that the three interior edges path is shorter than the exterior one, then of two triangles, also with the exterior edge greater than the concatenation of its two interior ones. Finally the closed configuration C_n possesses a closing cycle, made u of $n+2$ consecutive interior edges, whose concatenation is greater than the exterior path ω. We give an example of labeled O_4 corresponding to the reduction of $\beta_3\beta_2\gamma_2\gamma_3$:

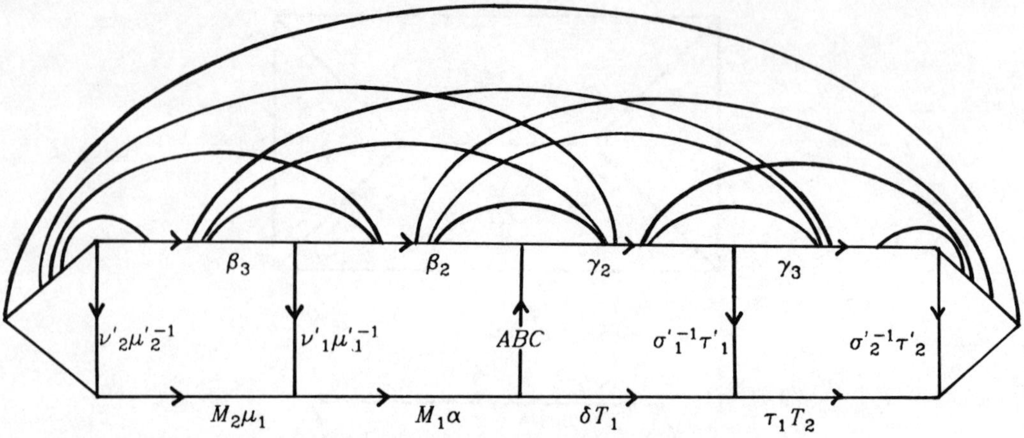

Fig. 5.8 : O_4.

All forbidden C_i, $i=0,\ldots,4$, are represented on this graph. Thus, our third assumption, which includes as a special case the Hypothesis 2, is:

Hypothesis 3 : In C_G, for all diagrams O_n, $n>2$, no associate diagram C_n exists.

Thus the three words $\nu_2\beta_3\beta_2\gamma$, $\nu_1\beta_2\gamma_2\sigma_1$ and $\beta\gamma_2\gamma_3\sigma_2$ are F,Γ-irreducible. Under the assumption that no C_n appears in C_G, this propagation of irreducible subwords goes on:

Lemma 5.14

The word Y_1 reduces on $Y_3=\nu_n\beta_{n-1}\cdots\beta_2\gamma_2\cdots\gamma_{m-1}\sigma_m$, with for $i=2\cdots n-1$ and $j=2\cdots m-1$:

i) $\beta_i\neq 1$ *and* $\gamma_j\neq 1$.

ii) *The word Y_3 is F,Γ-irreducible.*

Proof. By induction on $n+m$. The induction hypothesis is the structure of the previous rules and the irreducibility of the word still reduced:

$$\begin{cases} n_l\ ;\ \tau'_l\tau_l \longrightarrow \sigma'_l\gamma_l\sigma'^{-1}_{l+1}\tau'_{l+1}T^{-1}_{l+1} \\ m_k\ ;\ \mu'_k\mu_k \longrightarrow M^{-1}_{k+1}\mu'_{k+1}\nu'^{-1}_{k+1}\beta_k\nu'_k \end{cases}$$

where $\beta_k\neq 1$ and $\gamma_l\neq 1$, $\mu'_{k+1}\nu'^{-1}_{k+1}$ and $\sigma'^{-1}_{l+1}\tau'_{k+1}$ are pieces. The word $Y_{k,l}=\nu_k\beta_{k-1}\cdots\beta_2\gamma_2\cdots\gamma_{l-1}\sigma_l$ being irreducible. We have seen that this hypothesis holds for rules k,l,m_1 and n_1. And the next reductions must freely cancel either the word M_{k+1} or T_{l+1}. After that the reduction by either rule m_{k+1} or n_{l+1} may occur. The description of cases $\mu_2 M_2 \nu_1 \beta_2 \gamma$ and $\beta \gamma_2 \sigma_1 T_2 \tau_2$ shows that these two new rules have the previous form. Then, the word $Y_{k+1,l+1} = \nu_{k+1} \beta_k \cdots \beta_2 \gamma_2 \cdots \gamma_l \sigma_{l+1}$ is irreducible by the non-existence of the configurations C_i, $i=1 \cdots \max(k+1, l+1)$. Note that the deductions using the non-confluence of Y_j, j=1,2, are now replaced by the irreducibility of both $M_i \mu_{i-1}$ and $\tau_{j-1} T_j$ ▪

We have proved that the word Y_1 reduces on a non-null word Y_3, F,Γ-irreducible, in contradiction with the construction of Y. Therefore, under Hypothesis 2 (C'(4)) and 3 ($C_j, j=1 \cdots$), if a word W reduces on 1, then 1 is its only irreducible form:

$$\forall W,\ 1 \in Irred(W) \Rightarrow Irred(W) = \{1\} \tag{P}$$

Let W be a word such that $W =_G 1$. There exist words W' and $T_1, \ldots, T_k$ such that $W' = T_1 R_1 T_1^{-1} \cdots T_k R_k T_k^{-1}$, where $R_i \in S$, the ssr of defining relations. Moreover W is computed from W' by adjunction or deletion of F-redexes. We thus have the following diagram:

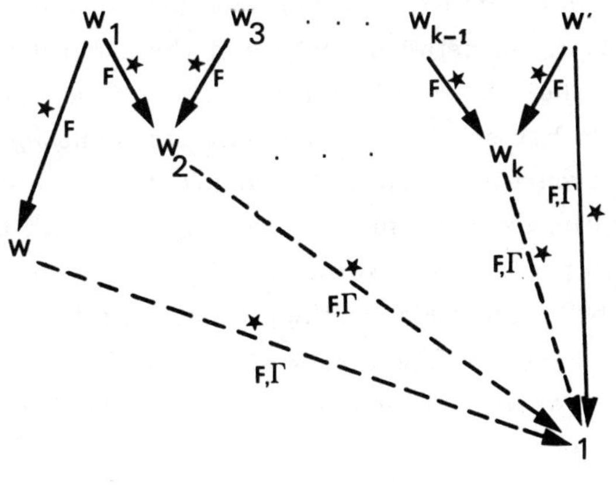

Fig. 5.9

which concludes the proof:

$$\forall W,\ W =_G 1 \Rightarrow Irred(W) = \{1\}.$$

Theorem 5.15

Let $\mathbf{G}=(G,S)$ *be a finite group presentation satisfying* $\mathbf{C}'(4)$ *and without S-diagram* $C_i, \forall i \in \mathbb{N}$, *for a length ordering. then the symmetrized presentation of* $\mathbf{G}$ *computed with this length ordering solves the word problem for* $\mathbf{G}$.

The fundamental result of small cancellation is a direct consequence of this theorem. As $\mathbf{G}$ satisfies $\mathbf{C}'(4)$, if $\mathbf{G}$ also satisfies $\mathbf{T}$, then its Cayley diagram does not possess any C_i configuration as they all include an interior vertex of degree three (cf. the end of §5.1). And if $\mathbf{G}$ satisfies $\mathbf{C}'(6)$, then by $\mathbf{C}'(4)$, C_0 is prohibited, while all the other ones include three consecutive pieces reducing to the rest of the elementary circuit or relation (i.e to the words $M_{i+1}\mu_i$, $M_1\alpha$, δT_1 and $\tau_j T_{j+1}$). This is impossible by $\mathbf{C}'(6)$ as no rule whose left member is composed of three consecutive pieces is allowed.

5.5 EXAMPLES

The present proof was initiated by H.Bücken [Büc79b]. He began an analysis of the symmetrization (lemma 5.2), isolated the words Y, and gave the irreducible form of Y_1. The present proof profits from experimentations done with our LISP system Al-Zebra [LeC84]. More precisely, a detailed analysis of the symmetrization was then available, enabling by the study of context reductions the outcome of configurations. This study has shown that the contexts, while indexed by $\mathbf{N}$, are finitely generated by the symmetrized presentation, also their non-existence can be checked. In contrast to the usual geometric proof [Lyn77,ch.5], the present one deals only with elementary properties of words and F cancellation, as does the proof by Greendlinger [Gre60a,b]. In consequence, the present proof is also longer.

Now, we may discuss the two hypotheses of the last theorem. There are two distinct hypotheses, first $\mathbf{C}'(4)$, second the non-existence of the C_i configurations. As the first assumption implies the non-existence of C_0, we may ask whether $\mathbf{C}'(4)$ is only used for this purpose? The answer is no, as shown by the following symmetrized set of the group $G_1=(a,b;ababa^{-1}b^{-1})$:

1:	$a^{-1}ba$	$\rightarrow$	bab
2:	bab^{-1}	$\rightarrow$	aba
3:	$a^{-1}b^{-1}a^{-1}$	$\rightarrow$	$ba^{-1}b^{-1}$
4:	$b^{-1}a^{-1}b^{-1}$	$\rightarrow$	$a^{-1}b^{-1}a$
5:	$b^{-1}ab$	$\rightarrow$	$ab^{-1}a^{-1}$
6:	$b^{-1}a^{-1}b$	$\rightarrow$	aba^{-1}
7:	$abab$	$\rightarrow$	ba

$$8: \quad aba^{-1}b^{-1} \rightarrow b^{-1}a^{-1}$$
$$9: \quad ba^{-1}b^{-1}a \rightarrow a^{-1}b^{-1}$$
$$10: \quad baba^{-1} \rightarrow a^{-1}b$$

The group is **C**'(2), but not **C**'(3). Hence it does not satisfy the first hypothesis. However, it satisfies **T** as every relation begins with one letter and ends with the other one. Thus no C_i configuration appears in its Cayley diagram. However, Dehn algorithm does not solve its word problem. Thus the condition **C**'(4) is necessary. If we take the same notations as in the proof, we have for example:

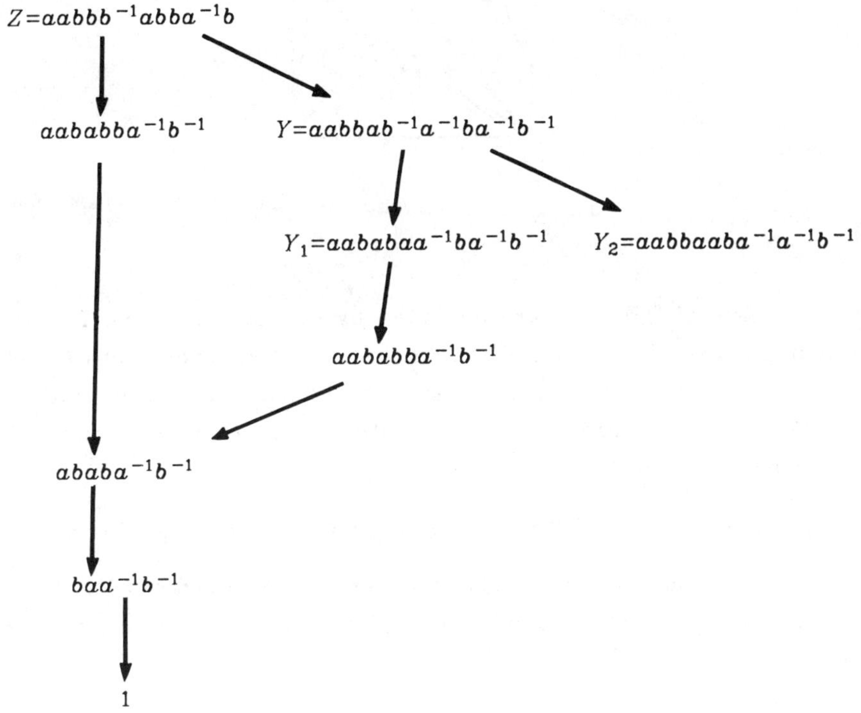

Fig. 5.10

The word Z both reduces to 1 and Y_2, irreducible. But Z is not minimal, Y is minimal (cf. beginning of prev. §) and the corresponding diagram is

$$\begin{cases} k : bab^{-1} \rightarrow aba \\ l : b^{-1}a^{-1}b \rightarrow aba^{-1} \\ m_1 : abab \rightarrow ba \\ m_2 : abab \rightarrow ba \end{cases}$$

The original piece is $b^{-1}a^{-1}$ with only $A=1$, then $\beta=ab=\mu_1=\mu'_1$. The point is that, no word $\nu_1\beta_2\gamma$ exists, as β is entirely cancelled by the m_1 reduction. The rule m_1 has its left member composed of the juxtaposition of two pieces, contradicting **C**'(4). The geometric interpretation has an interior vertex of degree four:

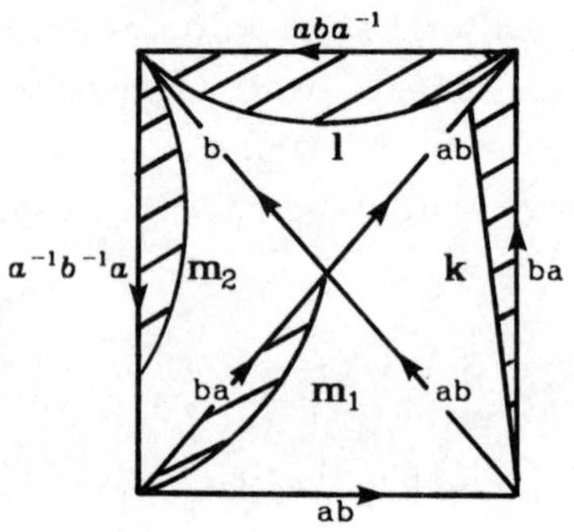

Fig. 5.11

The right members of the rules are delimited by the dashed areas. This example shows clearly our use of condition **C**'(4). The case $\lambda\rho\rightarrow\omega$, where both λ and ρ are pieces occurs with the rule **l**. The non-existence of such rules is constantly used in the study of critical pair reduction by a context. The other application of **C**'(4), including the previous one as special case, is:

$$\mathbf{C}'(4): \quad \text{if } P_1P_2U\in S, \text{ where } P_1,P_2 \text{ are pieces, then } U^{-1} \overset{*}{\underset{F.\Gamma}{\Rightarrow}} P_1P_2.$$

This proposition is an assumption on diagrams rather than a diagram itself.

Another example is the group $G_2=(A,B,C;ABC=CBA)$ whose one symmetrized set is:

1:	CBA	$\rightarrow$	ABC
2:	$A^{-1}CB$	$\rightarrow$	BCA^{-1}
3:	$B^{-1}A^{-1}C$	$\rightarrow$	$CA^{-1}B^{-1}$
4:	$C^{-1}B^{-1}A^{-1}$	$\rightarrow$	$A^{-1}B^{-1}C^{-1}$
5:	$B^{-1}C^{-1}A$	$\rightarrow$	$AC^{-1}B^{-1}$
6:	$C^{-1}AB$	$\rightarrow$	BAC^{-1}
7:	$ABCA^{-1}$	$\rightarrow$	CB
8:	$BCA^{-1}B^{-1}$	$\rightarrow$	$A^{-1}C$
9:	$CA^{-1}B^{-1}C^{-1}$	$\rightarrow$	$B^{-1}A^{-1}$
10:	$BAC^{-1}B^{-1}$	$\rightarrow$	$C^{-1}A$

This group is $\mathbf{C}'(5)$ but not $\mathbf{C}'(6)$. And it does not satisfy $\mathbf{T}$. Then, we have for example the following non-confluent derivations:

$$ABABABCA^{-1}CBCA^{-1}B^{-1}A^{-1}B^{-1}C^{-1}B^{-1}C^{-1}B^{-1}C^{-1}$$

$$\swarrow \qquad\qquad \searrow$$

$$ABABABCBCA^{-1}CA^{-1}B^{-1}AB^{-1}C^{-1}B^{-1}C^{-1}B^{-1}C^{-1}$$

$$ABABABCA^{-1}CA^{-1}CA^{-1}B^{-1}C^{-1}B^{-1}C^{-1}B^{-1}C^{-1}$$
$$\downarrow$$
$$ABABABCA^{-1}CA^{-1}B^{-1}A^{-1}B^{-1}C^{-1}B^{-1}C^{-1}$$
$$\downarrow$$
$$ABABCBCA^{-1}B^{-1}A^{-1}B^{-1}C^{-1}B^{-1}C^{-1}$$
$$\downarrow$$
$$ABABCA^{-1}CA^{-1}B^{-1}C^{-1}B^{-1}C^{-1}$$
$$\downarrow$$
$$ABABCA^{-1}B^{-1}A^{-1}B^{-1}C^{-1}$$
$$\downarrow$$
$$ABCA^{-1}B^{-1}C^{-1}$$
$$\downarrow$$
$$CBB^{-1}C^{-1}$$
$$\downarrow$$
$$1$$

Fig. 5.12

The diagram associated to this non-confluence is a five-rules one or a C_2 diagram:

$$\begin{cases} k : A^{-1}CB & \rightarrow BCA^{-1} \\ l : BCA^{-1}B^{-1} & \rightarrow A^{-1}C \\ m_1 : ABCA^{-1} & \rightarrow CB \\ n_1 : CA^{-1}B^{-1}C^{-1} & \rightarrow B^{-1}A^{-1} \\ o : BCA^{-1}B^{-1} & \rightarrow A^{-1}C \end{cases}$$

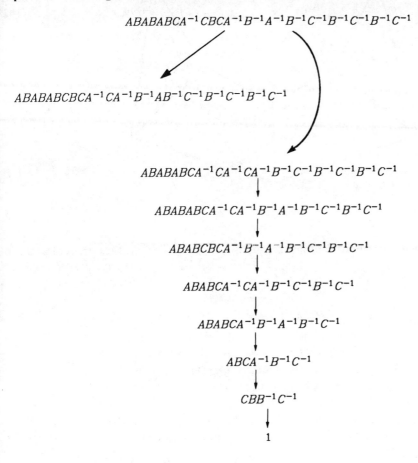

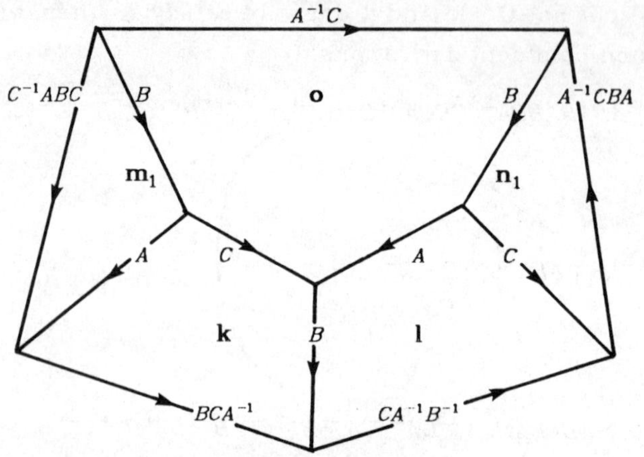

Fig. 5.13

However, the following symmetrized set for G_2 is also canonical: all its critical pairs are resolved.

1:	ABC	$\rightarrow$	CBA
2:	$C^{-1}B^{-1}A^{-1}$	$\rightarrow$	$A^{-1}B^{-1}C^{-1}$
3:	BAC^{-1}	$\rightarrow$	$C^{-1}AB$
4:	$CA^{-1}B^{-1}$	$\rightarrow$	$B^{-1}A^{-1}C$
5:	$A^{-1}CB$	$\rightarrow$	BCA^{-1}
6:	$B^{-1}C^{-1}A$	$\rightarrow$	$AC^{-1}B^{-1}$

The termination of this system is proved in the next chapter. The fact that the present proof applies to this symmetrized set follows from the ordering which is non-length-increasing, from the fact that all superpositions occur on non-pieces, thus no C_i diagram exists, and as pieces have length 1 while the relation has length 6, we trivially have $U \overset{*}{\underset{F,\Gamma}{\Rightarrow}} P_1P_2$ as required above. See [Ott84b] for another complete length-increasing system for this group.

Now, we may ask if some light has been brought on the geometrical nature of small cancellation groups. In any case, we got negative results telling what are the prohibited diagrams of their Cayley diagrams. First of all, we use an ordering on the paths (or on the words), verifying the conditions enumerated at the beginning of §5.2. Then our conditions are stated on both cuts done on the elementary circuits by this ordering (or rules cuts in left and right members), and diagrams extracted from the Cayley diagram. These conditions are:

i) For every elementary circuit, if P_1 and P_2 are two consecutive pieces and Q the rest of the circuit such that $P_1P_2Q^{-1} \in S$, then we must not have $P_1P_2 > Q$.

ii) For every open configuration $O_i, i \in \mathbb{N}$, such that its four-gons satisfy $P_iP'_iP''_i > Q$, the closing elementary circuit must not exist.

In conclusion, a small cancellation group is a **C'**(4) group such that there exists a length ordering for which no C_n, n>0, exist.

6 Complete group presentations

Rewriting systems have proved to be very useful in some varieties, such as multivariate commutative polynomials by the pioneer work of B. Buchberger. However, no systematic completions were done for groups despite numerous well-known families of group presentations. This chapter is a catalogue of complete systems for some classical groups, including surface, symmetric, polyhedral and Coxeter groups. These complete presentations possess interesting combinatorial properties: they succinctly describe Cayley graphs and provide efficient word problem algorithms. In general, when such a system exists, many others can be found. Also, we tried to present those with a small number of rules, or with an interesting proof of termination. The defining presentations are taken from [Cox72]. The complete presentations were intuited with a system Al-Zebra originally written in Maclisp on a Multics DPS-68, (now also available in Franz-LISP under Unix).

It was of interest to check whether or not the concrete groups were tractable by our limited rewriting means. The answer seems to be affirmative and to illustrate the fact that a general negative result, the undecidability of the word problem, must not be taken as a negative one for concrete problems. The catalogue length must not hide the complexity of completion. In groups, this procedure is related to the shortest path problem in graph theory, or to the enumeration of elementary circuits, which are hard problems. However, this algorithm is a compilation process, and obviously one is ready to pay a high price once for all in such computations. Furthermore, we now have three levels for data-structures: the first order terms, the monoid one, and this chapter details some cases of the last one, when a parametrised complete system may be compiled into stacks and pointers implementing an efficient word problem algorithm. This last stage is deduced by the same ideas of specialization for efficiency used to find the second one.

The reader will find complete presentations of:

- The seventeen two-dimensional space groups of translations, rotations and reflections of the euclidean plane which covers this plane by the transforms of a fundamental region.

- The Coxeter groups, generated by reflections through hyperplanes. Here, the partial commutativity of some generators is a case of failure. More interesting is the fact that this class possesses infinite sets of rules. Consequently, the full proof of correctness for the complete presentations is given.
- The polyhedral groups or groups generated by rotations in the spherical or hyperbolic geometry paving the whole sphere or half-plane by non-regular polygons.
- The fundamental groups of p-hole torus and projective planes. They initiated Dehn's study of small cancellation. The complete presentations have nice combinatorial properties.
- The symmetric groups S_n. Several complete systems exist, depending on the given presentations. They also have nice properties. Generally these systems have a great number of rules (sometimes $n!$, the size of S_n), but far from the bound given in § 3.2, as we choose presentations with numerous generators. However, the rules may be parametrized nicely, and the rules of a complete presentation of S_n belong to S_{n+1}. The presentations are closely related to some sorting algorithms.

6.1 THE SEVENTEEN PLANE GROUPS

We begin with some simple presentations. They are the celebrated geometric groups that cover the plane from a basic pavement. All of them have complete presentations; some of them give interesting termination problems. The first one is **p1** in the notation of [Cox72], and is defined by **p1**$=(X,Y;XY=YX)$ where X and Y are two non-collinear translations. Of course **p1** is just the commutative group on two generators. However, we give the complete presentation whose rules frequently appear in other complete systems, when two generators commute:

$$\mathbf{p1}\left\{\begin{array}{lcl} YX & \rightarrow & XY \\ YX^{-1} & \rightarrow & X^{-1}Y \\ Y^{-1}X & \rightarrow & XY^{-1} \\ Y^{-1}X^{-1} & \rightarrow & X^{-1}Y^{-1} \end{array}\right.$$

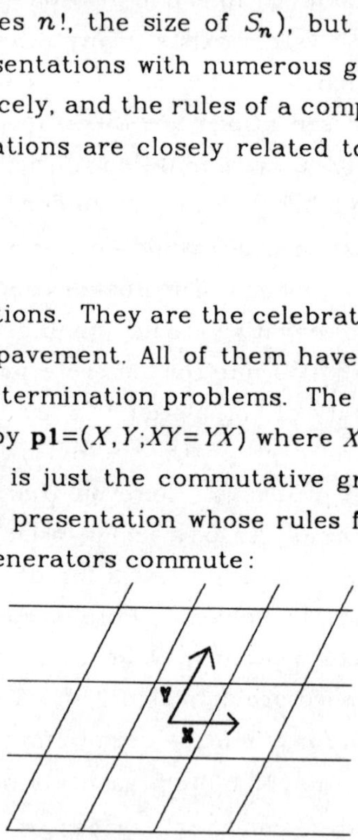

Termination: Lex. Ordering. with $Y^{-1}>Y>X^{-1}>X$.

The group is $\mathbf{Z}\times\mathbf{Z}$; the canonical forms are $([X+X^{-1}])^*([Y+Y^{-1}])^*$.

We follow Coxeter and Moser in their enumeration of the seventeen groups by adding new generators. Thus, let T be a rotation of angle π which maps the translations X and Y on their inverses, we get the group **p2** defined by the equations

$XY=YX,\ T^2=1,\ TXT=X^{-1},\ TYT=Y^{-1}.$

$$\mathbf{p2}\left\{\begin{array}{lcl} T^{-1} & \rightarrow & T \\ TY^{-1} & \rightarrow & YT \\ TX^{-1} & \rightarrow & XT \\ X^{-1}Y & \rightarrow & YX^{-1} \\ T^2 & \rightarrow & 1 \\ XY & \rightarrow & YX \\ X^{-1}Y^{-1} & \rightarrow & Y^{-1}X^{-1} \\ XY^{-1} & \rightarrow & Y^{-1}X \\ TX & \rightarrow & X^{-1}T \\ TY & \rightarrow & X^{-1}T \end{array}\right.$$

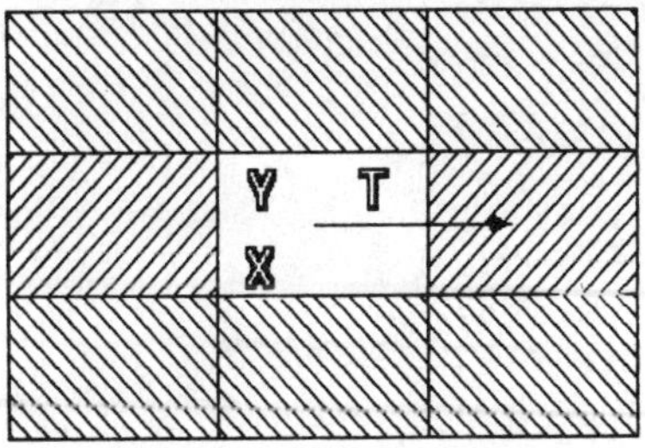

Termination: Lex Ordering with $T^{-1}>T>Y^{-1}>Y>X>X^{-1}$.

The canonical forms are $([Y+Y^{-1}])^{*}([X+X^{-1}])^{*}[T]$. Geometrically, if X denotes a translation along the X-axis and Y along the Y-axis of coordinates, a region is determined by its two coordinates equal to the number of each translation, and a sign + or - if the picture in the region is half-turned or not. Note that the two previous systems are symmetrized ones: no new consequence is added.

When the two translations are orthogonal, a reflection T parallel to Y's direction defines the group **pm2**$=(X,Y,T;XY=YX,TT^2=1,TYT=Y,TXT=X^{-1})$

$$\mathbf{pm}\left\{\begin{array}{lcl} X^{-1} & \rightarrow & TXT \\ T^2 & \rightarrow & 1 \\ XY & \rightarrow & YX \\ T^{-1} & \rightarrow & T \\ XY^{-1} & \rightarrow & Y^{-1}X \\ XTX & \rightarrow & T \\ TY & \rightarrow & YT \\ TY^{-1} & \rightarrow & Y^{-1}T \end{array}\right.$$

Termination: KB ordering with $\pi(X)=4$, $\pi(T)=\pi(T^{-1})=\pi(X)=\pi(Y)=\pi(Y^{-1})=1$ and $T^{-1}>T>X>Y>Y^{-1}>X^{-1}$.

The rules are not length-preserving ones as $X^{-1}\rightarrow TXT$, we may keep X's inverse to get the system:

$$\mathbf{pm}\begin{cases} T^{-1} & \rightarrow & T \\ T^2 & \rightarrow & 1 \\ YX & \rightarrow & XY \\ YX^{-1} & \rightarrow & X^{-1}Y \\ Y^{-1}X^{-1} & \rightarrow & X^{-1}Y^{-1} \\ Y^{-1}X & \rightarrow & XY^{-1} \\ TX & \rightarrow & X^{-1}T \\ TX^{-1} & \rightarrow & XT \\ TY & \rightarrow & YT \\ TY^{-1} & \rightarrow & Y^{-1}T \end{cases}$$

Termination: Lex Ordering with $T^{-1}>T>Y^{-1}>Y>X^{-1}>X$.

The canonical form is that of **p2**, but geometrically the picture in the negative regions is the mirror one of the positive regions, not the half-turned one.

Still in the case when X and Y are orthogonal, we may add to **p1** a transformation P, product of a reflection parallel to Y and a translation of square equal to Y. The group is called **pg**. As $Y=P^2$, putting $Q=PX$, we can eliminate X any Y to find that $\mathbf{pg}=(P,Q;P^2=Q^2)$.

$$\mathbf{pg}\begin{cases} P^{-1} & \rightarrow & Q^{-2}P \\ P^2 & \rightarrow & Q^2 \\ PQ^{-2} & \rightarrow & Q^{-2}P \\ Q^{-1}PQ & \rightarrow & QPQ^{-1} \\ Q^2P & \rightarrow & PQ^2 \end{cases}$$

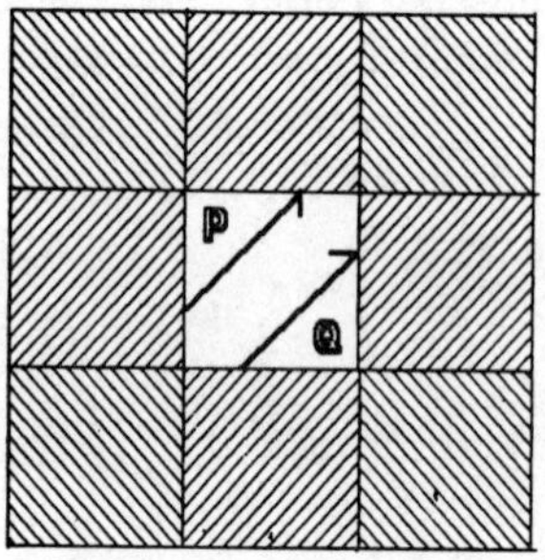

The two classical orderings, RPO and KB, fail to prove the termination as we must have $P>Q^{-1}>Q>P$. The observations in §3.2 give the following statements:

- We may restrict ourselves to reduction chains not including P^{-1}.
- Then, the number of P decreases or is constant in any reduction step. Therefore, this number may be supposed to be constant and the derivations act on words of the type $w_0Pw_1 \cdots w_{n-1}Pw_n$, $w_i \in \{Q,Q^{-1}\}^*$ and w_i not empty for $i=1,\ldots,n-1$.
- The same idea implies that the reductions $QQ^{-1}\rightarrow 1$ and $Q^{-1}Q\rightarrow 1$ are not used after some step. Then, the reductions act on words of constant length by the last three rules. To conclude the proof, we use an induction on this length n, trivially the rewriting halts when $n=3$. Now, either the first letter is never rewritten, in which case the induction hypothesis concludes the proof, or this letter is reduced at least once. Let us show that it is reduced at most once. We first observe that

the letters P,Q,Q^{-1} appear in only one rule as first letter. Then the reduction $PQQ \overset{*}{\to} QQP$ necessarily rewrites the first P. But this is possible only with the rule $PQ^{-1}Q^{-1} \to Q^{-1}Q^{-1}P$, thus we must have the reduction $QQ \overset{*}{\to} Q^{-1}Q^{-1}$. In turn this is possible only with $QQP \to PQQ$. But this step does not yield a Q^{-1} in first position, and we must use another rewriting at first index which is impossible by induction hypothesis. The remaining cases for the right member $Q^{-1}Q^{-1}P$ being reduced in the last two left members uses the same reduction $QQ \overset{*}{\to} Q^{-1}Q^{-1}$ at the previous induction level. And the same fact appears for the two other right members:
QPQ^{-1} needs the rewriting of $PQ^{-1} \overset{*}{\to} QP$ which is impossible.
$Q^{-1}Q^{-1}P$ needs $QP \overset{*}{\to} PQ^{-1}$, also impossible.
Thus the system **pg** is noetherian.

We will use this reasoning once more, in the section devoted to surface groups. Then, for the group **pg**, the irreducible words are:

$$[P](QP)^*(Q^{-1}P)^*[Q^{-1}] + (Q^{-1})^*(PQ^{-1})^*[P] + [P](QP)^*Q^*$$

Permuting the generators P and Q with a reflection R, we define the group **cm**$=P^2=Q^2,R^2=1,RPR=Q$. The generator Q may be eliminated, we get the complete presentation

$$\mathbf{cm}\begin{cases} R^{-1} \to R \\ RP \to P^2RP^{-1} \\ RR \to 1 \\ RP^{-2} \to P^{-2}R \end{cases}$$

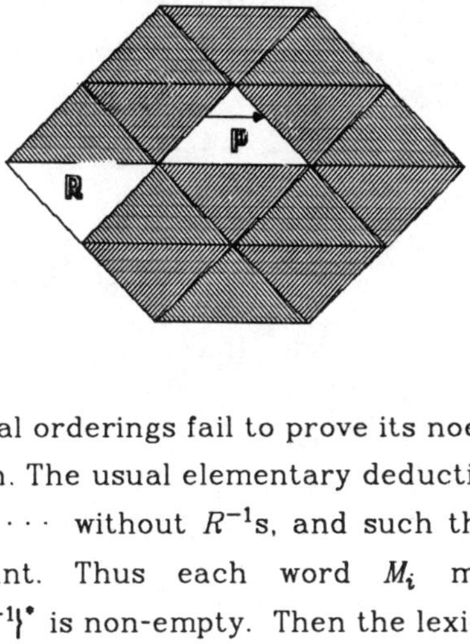

As for the previous example, the usual orderings fail to prove its noetherianity. Note that the rules do not preserve length. The usual elementary deductions yield reduction chains $M_1 \to M_2 \to \cdots \to M_n \to \cdots$ without R^{-1}s, and such that $n(M_i,R)$, the number of R in M_i, is constant. Thus each word M_i may be written $w_o R w_1 \cdots w_{n-1} R w_n$, where $w_j \in \{P,P^{-1}\}^*$ is non-empty. Then the lexicographical ordering on the $2n$-tuples of integers
$<n(w_n,P),\ldots,n(w_0,P),n(w_n,P^{-1}),\ldots,n(w_1,P^{-1})>$
proves the termination as rules 2 and 4, together with $PP^{-1} \to 1$ and $P^{-1}P \to 1$ strictly decrease these $2n$-tuples. The canonical form of the words is $(P^*+(P^{-1})^*)(RP^{-1})^*[R]$.

The group **pm**, with a reflection defines the group **pmm**, whose one presentation is $A^2=B^2=C^2=D^2=(AB)^2=(BC)^2=(CD)^2=(DA)^2=1$. The completion gives the fourteen rules system:

$$\mathbf{pmm}\left\{\begin{array}{lcl} A^{-1} & \rightarrow & A \\ B^{-1} & \rightarrow & B \\ C^{-1} & \rightarrow & C \\ D^{-1} & \rightarrow & D \\ AA & \rightarrow & 1 \\ BB & \rightarrow & 1 \\ CC & \rightarrow & 1 \\ DD & \rightarrow & 1 \\ BA & \rightarrow & AB \\ CB & \rightarrow & BC \\ DC & \rightarrow & CD \\ DA & \rightarrow & AD \\ CDB & \rightarrow & DBC \\ BCA & \rightarrow & CAB \end{array}\right.$$

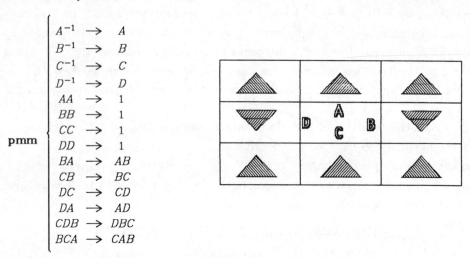

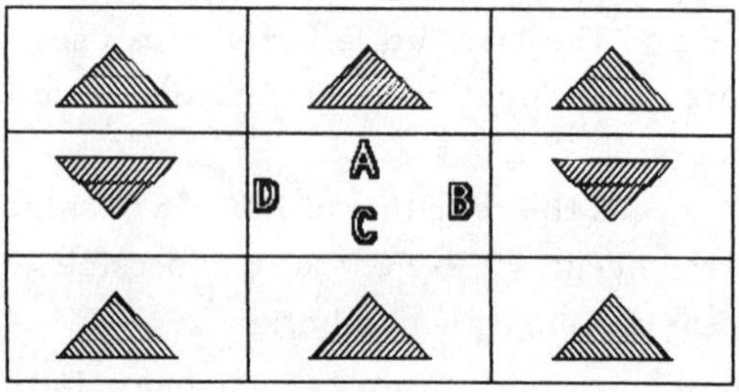

The termination results from a direct observation of the generator moves. From an infinite reduction chain, the classical length argument tells us that from some range of rewriting, only the last six rules are used. Therefore, the generator A must appear finitely many times in the remaining reductions as all the three rules where it appears move it leftwards. Only three rules remain, then B moves leftwards. Finally we are left with only one permutation rule which concludes the proof.

If we point out that the generators A and C,B and D do not commute, then we may enumerate the canonical forms:

$$[A](CA)^*((DB)^*[C][D] + (BD)^*[B[C[D]]]) + [C](AC)^*[D]$$

Adjoining a reflection R that reverses the directions of P an Q in **pg**, we define **pmg**, putting $A=R,B=PR$ and $C=QR$, we have the presentation $A^2=B^2=C^2=1$, $BAB=CAC$ whose one complete system is:

$$
\textbf{pmg}\left\{\begin{array}{ccc}
A^{-1} & \rightarrow & A \\
B^{-1} & \rightarrow & B \\
C^{-1} & \rightarrow & C \\
AA & \rightarrow & 1 \\
BB & \rightarrow & 1 \\
CC & \rightarrow & 1 \\
CBA & \rightarrow & ACB \\
BAB & \rightarrow & CAC \\
BCA & \rightarrow & ABC \\
CACB & \rightarrow & BA \\
CACAB & \rightarrow & BACAC
\end{array}\right.
$$

Termination: Lex Ordering with $B^{-1}>C^{-1}>A^{-1}>B>C>A$.

If the previous reflection R defining **pmg** is replaced by a half-turn T, we get **pgg** defined by $P^2=Q^2, T^2=1, TPT=Q$. We introduce the generator $O=PT$ and eliminate Q, so that a simpler definition is $(PO)^2=(P^{-1}O)^2=1$.

$$
\textbf{pgg}\left\{\begin{array}{ccc}
PO^{-1} & \rightarrow & OP^{-1} \\
PO & \rightarrow & O^{-1}P^{-1} \\
P^{-1}O^{-1} & \rightarrow & OP \\
P^{-1}O & \rightarrow & O^{-1}P
\end{array}\right.
$$

Termination: trivial.

In the same way, a half-turn added to **pmm**, permuting A and C, B and D, defines **cmm**, whose a presentation is $A^2=B^2=C^2=(AC)^2=(ABCB)^2=1$.

$$
\textbf{cmm}\left\{\begin{array}{ccc}
A^{-1} & \rightarrow & A \\
B^{-1} & \rightarrow & B \\
C^{-1} & \rightarrow & C \\
AA & \rightarrow & 1 \\
BB & \rightarrow & 1 \\
CC & \rightarrow & 1 \\
CA & \rightarrow & AC \\
BCBA & \rightarrow & ABCB \\
BABC & \rightarrow & CBAB \\
BABAC & \rightarrow & CBABA
\end{array}\right.
$$

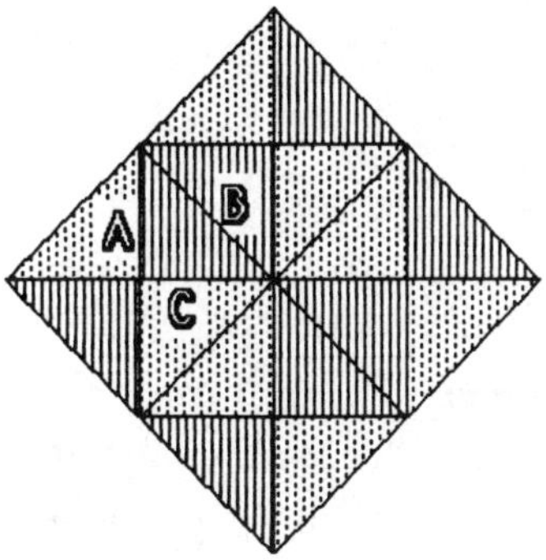

Termination: Lex Ordering with $B^{-1}>C^{-1}>A^{-1}>B>C>A$.

Coming back to **p2**, when its pavement is a square, we get the group **p4**, defined by two rotations A and B, of angle $\pi/4$, and the relations $A^4=B^4=(AB)^2=1$.

p4	$A^{-1}B^{-1}$	→	BA
	$B^{-1}B^{-1}$	→	BB
	$A^{-1}A^{-1}$	→	AA
	$B^{-1}A^{-1}$	→	AB
	ABA	→	B^{-1}
	AAA	→	A^{-1}
	BBB	→	B^{-1}
	BAB	→	A^{-1}
	$A^{-1}BA$	→	AAB^{-1}
	$A^{-1}BB$	→	BAB^{-1}
	$B^{-1}AB$	→	BBA^{-1}
	$B^{-1}AA$	→	ABA^{-1}

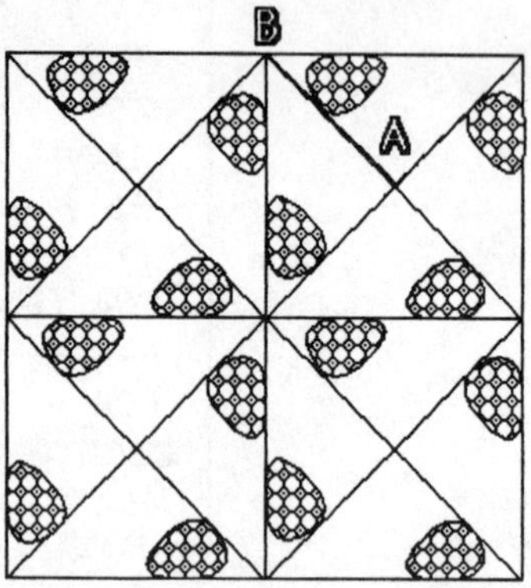

Termination: Lex Ordering with $A^{-1}>B^{-1}>A>B$.
The canonical forms are $[A+(A+B)A](B^{-1}A)^{*} + [B+(B+A)B](A^{-1}B)^{*}$.

When **pmm** also has a square as pavement, the number of possible transformations is incremented by a reflection, defining the group **p4m** presented by $A^2=B^2=C^2=(AB)^4=(BC)^2=(CA)^4=1$.

p4m	A^{-1}	→	A
	B^{-1}	→	B
	C^{-1}	→	C
	AA	→	1
	BB	→	1
	CC	→	1
	CB	→	BC
	$CACA$	→	$ACAC$
	$BABA$	→	$ABAB$
	$CABAB$	→	$BCABA$
	$CABCABA$	→	$ACABCAB$

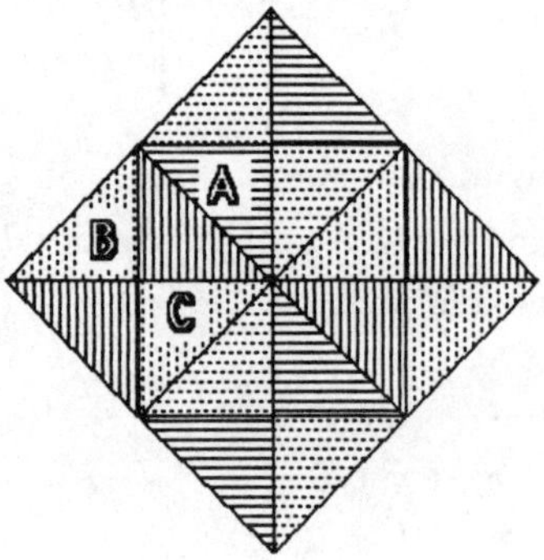

Termination: Lex Ordering with $C^{-1}>B^{-1}>A^{-1}>C>B>A$.

Always from **pmm**, we introduce a rotation B of $\pi/2$ that cyclically permutes the reflections of **pmm**. This defines **p4g**$=(A,B;A^2,B^4,(AB^{-1}AB)^2)$, the generators A and B being two of the reflections from **pmm**.

p4g	A^{-1}	→	A
	B^{-1}	→	BBB
	AA	→	1
	$BBBB$	→	1
	$ABBBAB$	→	$BBBABA$
	$ABABBB$	→	$BABBBA$
	$ABABBAB$	→	$BABBABA$

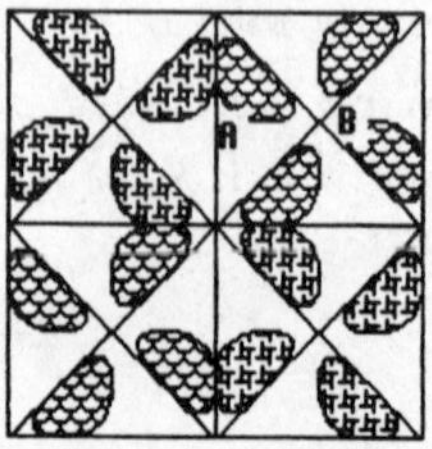

Termination: KB ordering with $\pi(B^{-1})=4, \pi(A^{-1})=\pi(A)=\pi(B)=1$ and $A^{-1}>B^{-1}>A>B$.

Coming back to **p1**, when the translations X and Y have the same length with their angle equal to $2\pi/3$, we add the rotation A which transforms Y into X. If $B=AX$ then we have the group **p3**, $A^3=B^3=(AB)^3=1$:

$$\mathbf{p3}\left\{\begin{array}{lcl} A^{-1} & \rightarrow & AA \\ B^{-1} & \rightarrow & BB \\ AAA & \rightarrow & 1 \\ BBB & \rightarrow & 1 \\ BABA & \rightarrow & AABB \\ BBAA & \rightarrow & ABAB \end{array}\right.$$

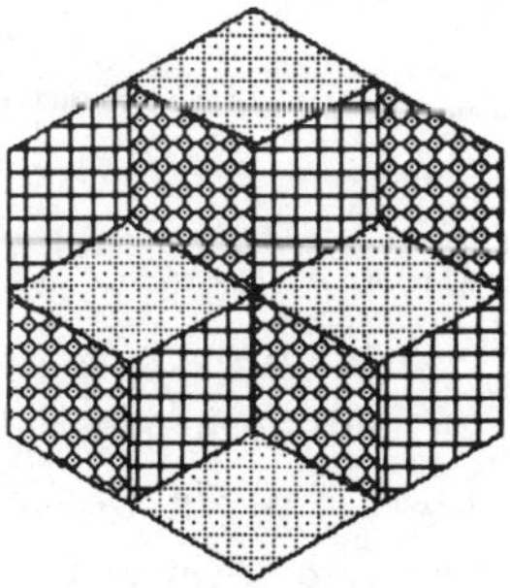

Termination: KB Ordering with $\pi(B^{-1})=\pi(A^{-1})=3$, $\pi(A)=\pi(B)=1$ and $B^{-1}>A^{-1}>B>A$. The normal forms are $[[A]A](BAA)^* + [[A]B](AAB)^*](BAB)^*[B[A]]$.

From **p3**, we define **p3m1** by a reflection mapping A on B^{-1}, whose a presentation is $C^2=B^3=(CB^{-1}CB)^3=1$.

$$\mathbf{p3m1}\left\{\begin{array}{ccc} C^{-1} & \rightarrow & C \\ CC & \rightarrow & 1 \\ BB & \rightarrow & B^{-1} \\ B^{-1}B^{-1} & \rightarrow & B \\ CB^{-1}CBC & \rightarrow & B^{-1}CBCB^{-1}CB \end{array}\right.$$

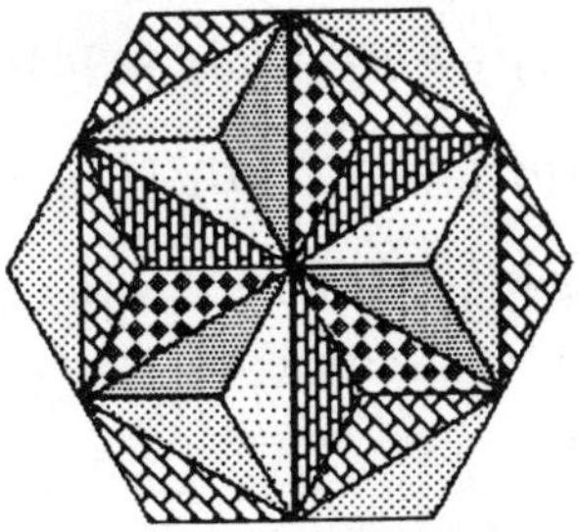

The proof of termination is difficult for this system. The last rule is such that every proper subword of its left member is also a proper subword of the right member. Thus all classical orderings based upon some decomposition of terms fail in such cases. Let us call R this last rule and G the remaining ones. Obviously, the relation $\overset{*}{\underset{G}{\rightarrow}}$ is noetherian, and its canonical forms are $[C]((B+B^{-1})C)^*[B+B^{-1}]$. In fact G is itself a canonical system. The reader may then check that on this language, the relation $\underset{R}{\rightarrow}$ is noetherian. This concludes the proof of noetherianity of the relation $\overset{*}{\underset{G}{\rightarrow}} \cdot \underset{R}{\rightarrow} \cdot \overset{*}{\underset{FG}{\rightarrow}}$.

Yet from **p3**, we introduce the group **p31m** by adding a reflection A that maps every rotation of **p3** on its inverse. We have the following definition with A and the product of A by the defining rotations of **P3** $A^2=B^2=C^2=(AB)^3=(BC)^3=(CA)^3=1$.

$$
\mathbf{p31m}\begin{cases}
A^{-1} \rightarrow A \\
B^{-1} \rightarrow B \\
C^{-1} \rightarrow C \\
AA \rightarrow 1 \\
BB \rightarrow 1 \\
CC \rightarrow 1 \\
CBC \rightarrow BCB \\
BAB \rightarrow ABA \\
ACA \rightarrow CAC
\end{cases}
$$

The proof of termination is the same as **pg**'s one. In an infinite chain of reductions, the length becomes stationary. Then only the last three rules may be used, and we prove that a rule can reduce a prefix in this chain at most once. This is obvious when the length is equal to three. In the general case, we may restrict our attention to a reduction of AU by symmetry of the rules. Thus,
$AU \xrightarrow{*} ACAV \rightarrow CACV$, and the rewriting of the first C needs the reduction
$ACV \xrightarrow{*} BCW$, but this in turn needs two reductions at index one. This is impossible by induction hypothesis. The proof gives the detail of the various canonical forms:

$$[[B]C](ABC)^*(ABA)^*(CBA)^*[C]B]]$$

$$+[[C]A](BCA)^*(BCB)^*(ACB)^*[A]C]]$$

$$+[[A]B](CAB)^*(CAC)^*(BAC)^*[B]A]]$$

A last group is defined from **p3** by the permutation of its rotations by a new rotation of angle π. With two rotations A $(2\pi/3)$ and B $(\pi/3)$, this group **p6** is defined by $A^3=B^6=(AB)^2=1$.

$$
\left\{
\begin{array}{rcl}
A^{-1}B^{-1} & \rightarrow & BA \\
A^{-2} & \rightarrow & A \\
A^{2} & \rightarrow & A^{-1} \\
B^{-1}A^{-1} & \rightarrow & AB \\
ABA^{-1} & \rightarrow & B^{-1}A \\
BAB & \rightarrow & A^{-1} \\
ABA & \rightarrow & B^{-1} \\
B^{-3} & \rightarrow & B^{3} \\
A^{-1}BA & \rightarrow & AB^{-1} \\
A^{-1}BA^{-1} & \rightarrow & AB^{-1}A \\
A^{-1}B^{3} & \rightarrow & BAB^{-2} \\
A^{-1}B^{2}A & \rightarrow & AB^{-1}AB^{-1} \\
AB^{2}A & \rightarrow & B^{-1}AB^{-1} \\
B^{4} & \rightarrow & B^{-2} \\
B^{-1}AB^{-1}A & \rightarrow & AB^{2}A^{-1} \\
B^{-2}AB & \rightarrow & B^{3}A^{-1}
\end{array}
\right.
$$

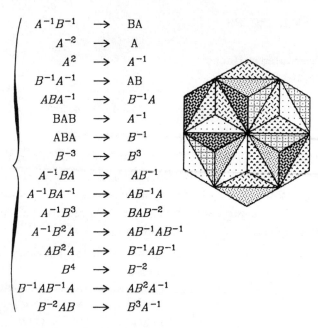

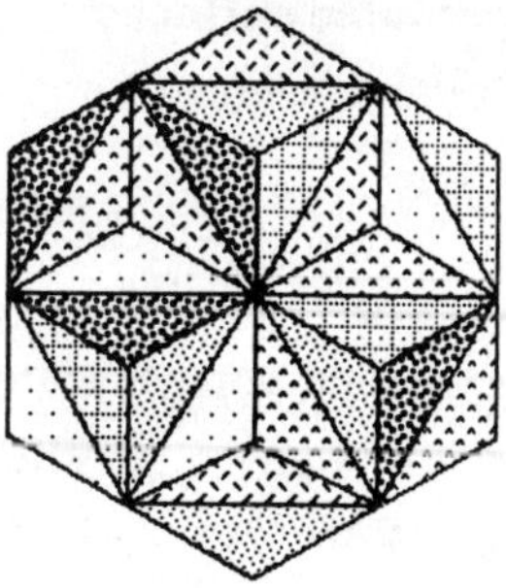

Termination: Lex Ordering with $A^{-1}>B^{-1}>A>B$.

The seventeenth group **p6m** is defined from **p31m** by a reflection that exchanges A and C, B being unchanged. A presentation for **p6m** is then $A^2=B^2=C^2=(AB)^2=(BC)^2=(CA)^6=1$.

$$
\mathbf{p6m}\left\{
\begin{array}{rcl}
A^{-1} & \rightarrow & A \\
B^{-1} & \rightarrow & B \\
C^{-1} & \rightarrow & C \\
AA & \rightarrow & 1 \\
BB & \rightarrow & 1 \\
CC & \rightarrow & 1 \\
BC & \rightarrow & CB \\
BAB & \rightarrow & ABA \\
BACB & \rightarrow & ABAC \\
CACACA & \rightarrow & ACACAC \\
BACABA & \rightarrow & ABACAB \\
BACACAC & \rightarrow & CBACACA \\
BACACBACACA & \rightarrow & ABACACBACAC
\end{array}
\right.
$$

Termination: Lex Ordering with $A^{-1}>B^{-1}>C^{-1}>B>C>A$.

6.2 SURFACE GROUPS

6.2.1 Orientable surfaces

The defining presentation of a p holes torus is

$$T_p=(A_1,\ldots,A_{2p};A_1\cdots A_{2p}=A_{2p}\cdots A_1).$$

The pieces of the symmetrized set of relations are just the generators and their inverses, so that when $p\geq 2$, the presentation is **C'**(6) and the family has a word problem solved by Dehn algorithm. Surprisingly, the Knuth-Bendix completion shows that Dehn algorithm not only insures the confluence on the unit element 1, but also gives a canonical form for each element in the present case, as the systems are at the same time symmetrized ones and complete ones.

The case $p=1$ defines the group **p1** of the previous section. The following case $p=2$ has been used as example in §5.2:

$$T_2\left\{\begin{array}{ccc}
DCBA & \to & ABCD\\
BCDA^{-1} & \to & A^{-1}DCB\\
B^{-1}A^{-1}DC & \to & CDA^{-1}B^{-1}\\
DA^{-1}B^{-1}C^{-1} & \to & C^{-1}B^{-1}A^{-1}D\\
D^{-1}C^{-1}B^{-1}A^{-1} & \to & A^{-1}B^{-1}C^{-1}D^{-1}\\
B^{-1}C^{-1}D^{-1}A & \to & AD^{-1}C^{-1}B^{-1}\\
BAD^{-1}C^{-1} & \to & C^{-1}D^{-1}AB\\
D^{-1}ABC & \to & CBAD^{-1}
\end{array}\right.$$

The second important result about this system, after the fact that it is both symmetrized and complete, is that all the rules have the form $\lambda\to\bar{\lambda}$, where $\bar{\lambda}$ is the reverse word of λ. In the general case, the completion gives the system T_p of $4p$ rules composed of words whose length is $2p$.

$$\left\{\begin{array}{ccc}
A_{2k}\cdots A_{2p}A_1^{-1}\cdots A_{2k-1}^{-1} & \to & A_{2k-1}^{-1}\cdots A_1^{-1}A_{2p}\cdots A_{2k}\\
A_{2k}\cdots A_1A_{2p}^{-1}\cdots A_{2k+1}^{-1} & \to & A_{2k+1}^{-1}\cdots A_{2p}^{-1}A_1\cdots A_{2k}\\
A_{2k}^{-1}\cdots A_1^{-1}A_{2p}\cdots A_{2k+1} & \to & A_{2k+1}\cdots A_{2p}A_1^{-1}\cdots A_{2k}^{-1}\\
A_{2k}^{-1}\cdots A_{2p}^{-1}A_1\cdots A_{2k-1} & \to & A_{2k-1}\cdots A_1A_{2p}^{-1}\cdots A_{2k}^{-1}
\end{array}\right.$$

where $k=1,\ldots,p$. *Termination*: Lex ordering such that
$A_{2p}>A_{2p}^{-1}>A_{2p-2}>A_{2p-2}^{-1}>\cdots>A_2>A_2^{-1}>A_1>A_1^{-1}>A_3>A_3^{-1}>\cdots>A_{2p-1}>A_{2p-1}^{-1}$.

Now, with this family of rewriting rules, we have reached a new step in our aim of designing new algorithms from a general one acting on a special kind of data: the sets T_p can be specialized in new data structures and algorithms. The first point is to find an adequate representation for words. Then the rewriting sets define three new algorithms: one for reducing an arbitrary ***free*** element of this structure to an irreducible one, and two other ones to perform efficiently the two basic operations in groups on reduced objects, the multiplication and the computation of the inverses. An obvious solution for the former operation is to concatenate two irreducible words, and then to reduce this new word. This algorithm is inefficient as it does not use the fact that a reduction may occur only at the joint of the two initial words.

It is out of the scope of this monograph to detail such algorithms. Let us just mention a first insight about an efficient computation of the normal forms by giving an upper bound to the number of T_p-reductions. Book [Boo82a] has proved that without length-increasing rules, a rewriting system on words possess a linear-time algorithm computing normal forms. In the present case, such an algorithm does not need to perform backward search after a reduction. We first make two observations. If we do not distinguish between the generators and their inverses, the rules reduce to two distinct types. Let the letter a^k denote either the generator A_k or its inverse, with the obvious meaning for a_k^{-1}. Only even letters a_{2k} appear as first letter in rules left members. Let $M = Wa_{2k} \cdots a_{2k+1}^{-1} W'$ be a word whose leftmost T_p-redex is the one displayed (the other type of reduction ending with a_{2k-1} is similar). We assume that Wa_{2k} is F-reduced, then $M \rightarrow Wa_{2k+1}^{-1} \cdots a_{2k} W'$. What are the following possible F or T_p-reductions occurring on a suffix of W? We have two cases, either the suffix will be F or T_p-reduced:

1) F-reduction. This reduction is $a_{2k+1} a_{2k+1}^{-1} \rightarrow 1$, and $M \underset{T_p}{\longrightarrow} \cdot \underset{F}{\longrightarrow} W_0 a_{2k+2}^{-1} a_{2k+3}^{-1} \cdots a_{2k-1} a_{2k}$. No further reduction is then possible. For a F-redex implies $W = W_0 a_{2k+1} = W_1 a_{2k+2} a_{2k+1}$ and the initial T_p-redex would not be the leftmost one. Any T_p-redex implies $W = W_0 a_{2k+1} = W_1 a_{2k}^{-1} a_{2k+1}^{-1} a_{2k+1}$ and the subword preceding the initial redex would not be F-irreducible.
2) T_p-reduction. We still have $M \underset{T_p}{\longrightarrow} Wa_{2k+1}^{-1} a_{2k+2}^{-1} \cdots a_{2k} W'$. Then every T_p-redex using the suffix either must include the letter a_{2k}^{-1}, but this is impossible as the word $Wa_{2k} = W_0 a_{2k}^{-1} a_{2k}$ would be F-irreducible, or this new redex inverses its slope just on the joint of W and the initial redex, that is $W = W_1 a_{2j} a_{2j+1} \cdots a_{2p} a_1^{-1} \ldots a_{2p}^{-1}$, but once more the initial word would be F-reducible by $a_{2p} a_{2p}^{-1} \rightarrow 1$.

To resume, the leftmost redex strategy needs at most one F-reduction backwards after a T_p-reduction. Now, what is the next index to look at for the possible following reduction? If the following T_p-redex has a common subword with the right member just introduced, we are in the following case:

$$a_{2k+1}^{-1} \cdots a_{2j}^{-1} \cdots a_{2p}^{-1} a_1 \cdots a_{2k} a_{2k+1} \cdots a_{2j-1} W'_0 \text{ with } a_{2k+1} \cdots a_{2j-1} W'_0 = W'$$

but this implies that M would be F-reducible at the joint of W' and the initial redex. Thus, we deduct that we must T_p-reduce the leftmost redex only if we cannot F-reduce on both sides of this T_p-redex. The case is similar when the inversion of the redex slope occurs at the junction (cf. previous analysis). Thus we can restrict the search of left members on W' . But F-reductions are possible after the initial reduction with W'. This leads to a special case where the new right member disappears entirely:

$$M = Wa_{2k} \cdots a_1 a_{2p}^{-1} \cdots a_{2k+1}^{-1} a_{2k}^{-1} a_{2k-1}^{-1} \cdots a_1^{-1} a_{2p} \cdots a_{2k+1} W'_0 \rightarrow WW'_0$$

and in this case we must move backwards in W of $2p-1$ letters to restart the reductions. However, the previous special case can be checked easily, and it reduces the length of M by $4p$ letters, moving backwards of $2p-1$ after this is then as going forward by $2p+1$. So that in any case we have at most $|M|/2p$ T_p-reductions to reach the normal form of M, and this analysis sketches the reduction algorithm.

The group $(A,B,C;ABC=CBA)$ used in §5.4 has a symmetrized set that is also canonical.

$$\perp_1 \left\{ \begin{array}{rcl} CBA & \rightarrow & ABC \\ BCA^{-1} & \rightarrow & A^{-1}CB \\ B^{-1}A^{-1}C & \rightarrow & CA^{-1}B^{-1} \\ A^{-1}B^{-1}C^{-1} & \rightarrow & C^{-1}B^{-1}A^{-1} \\ AC^{-1}B^{-1} & \rightarrow & B^{-1}C^{-1}A \\ C^{-1}AB & \rightarrow & BAC^{-1} \end{array} \right.$$

But its noetherianity does not follow from a classical ordering. It belongs to the family $\perp_p$ defined by $(A_1, \ldots, A_{2p+1} ; A_1 \cdots A_{2p+1} = A_{2p+1} \cdots A_1)$. Here is a complete presentation $\perp_p$ with $4p+2$ rules. The words'length is $2p+1$.

$$\begin{cases} A_{2k+1} \cdots A_1 A_{2p+1}^{-1} \cdots A_{2k+2}^{-1} & \rightarrow A_{2k+2}^{-1} \cdots A_{2p+1}^{-1} A_1 \cdots A_{2k+1} \\ A_{2k+1}^{-1} \cdots A_{2p+1}^{-1} A_1 \cdots A_{2k} & \rightarrow A_{2k} \cdots A_1 A_{2p+1}^{-1} \cdots A_{2k+1}^{-1} \\ A_{2k} \cdots A_{2p+1} A_1^{-1} \cdots A_{2k-1}^{-1} & \rightarrow A_{2k-1}^{-1} \cdots A_1^{-1} A_{2p+1} \cdots A_{2k} \\ A_{2k}^{-1} \cdots A_1^{-1} A_{2p+1} \cdots A_{2k+1} & \rightarrow A_{2k+1} \cdots A_{2p+1} A_1^{-1} \cdots A_{2k}^{-1} \end{cases}$$

where $k=0, \ldots, p$. The noetherianity of $\perp_p$ follows from the fact that each member of $G \cup G^{-1}$ appears in one and only one rule as prefix. The rules having the same length, we may restrict our attention to reduction chains of words of the same length.

Lemma 6.1

Let U and V be two words of the same length, and $b_1 \cdots b_{2p}$ be a left member prefix, then the reduction $b_1 \cdots b_{2p} U \xrightarrow[\perp_p]{*} b_{2p}^{-1} \cdots b_1^{-1} V$ is impossible.

Proof. The proof is by induction on $|U|$. The proposition is trivial for both $|U|=0$ and $|U|=1$ from left members'irreducibility. In the general case, let $P=b_1 \cdots b_{2p}=A_{2p+1} \cdots A_2$, the prefix of the first rule ($k=0$ above), due to the symmetry of the rules the remaining cases are similar. Then any reduction from PU to $P^{-1}V$ uses the first rule
$A_{2p+1} \cdots A_1 \rightarrow A_1 \cdots A_{2p+1}$ as $b_1=A_{2p+1}$ must be reduced and this is the only rule starting with this letter (let us recall that reductions preserve length: we cannot have $A_{2p+1}A_{2p+1}^{-1} \rightarrow 1$). Then we have the following sequence of reductions:

$$b_1 \cdots b_{2p} U \xrightarrow{*} A_{2p+1} \cdots A_2 A_1 U'$$

$$\rightarrow A_1 A_2 \cdots A_{2p+1} U'$$

Afterwards, A_2^{-1} at index 1 is possible only when A_1 disappears, which in turn is possible only by rule $A_1 A_{2p+1}^{-1} \cdots A_2^{-1} \rightarrow A_2^{-1} \cdots A_{2p+1}^{-1} A_1$, thus we have necessarily $A_2 \cdots A_{2p+1} U' \xrightarrow{*} A_{2p+1}^{-1} \cdots A_2^{-1} V'$, which contradicts the induction hypothesis ∎
We have proved the proposition of §3.2.2:

Corollary 6.2

Let U and V be such that $|U|=|V|$, then one never has $\rho_i U \xrightarrow[\perp_p]{*} \lambda_j V$, ρ_i (resp. λ_j) right member of a rule (resp. left).

Thus the reductions must halt as a prefix may be reduced only once.

Geometrically, the two families of complete presentations possess a terse description with $4p$ and $4p+2$ gons. We give them for T_2 and $\perp_1$:

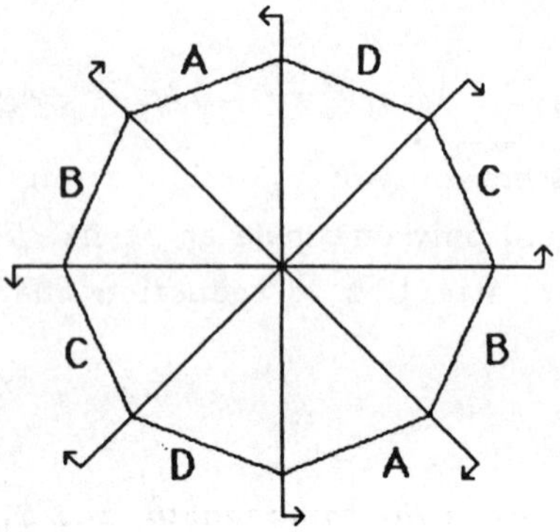

Fig. 6.1, T_2

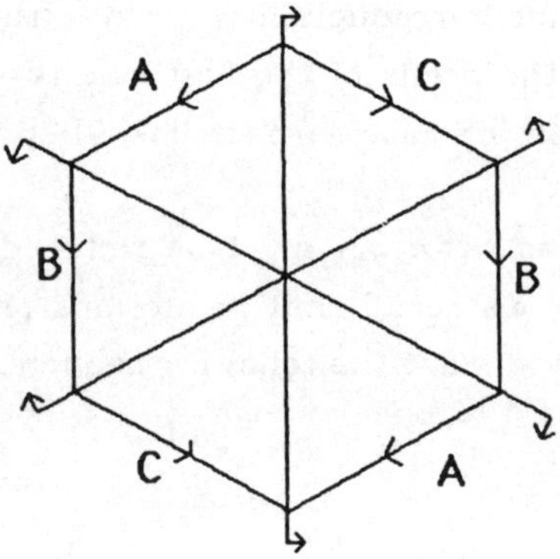

Fig. 6.2, $\perp_1$

The polygons represent an elementary circuit from the Cayley diagram of the groups. As the groups are defined by a single relation, all the relations from the ssr can be read on these pictures, in clockwise order and its opposite. Two paths start from a given vertex to the opposite one. The arrows show the paths selected as the irreducible ones by the completion algorithm: at a vertex, you must follow the arrow in order to reach the opposite side. All the rules are coded on these graphs. Observe that we are always in the second cases of both i) and ii) from lemma 5.3. These symmetrized sets, which are also canonical, illustrates very well the effect of the symmetrization on the Cayley diagram. It cuts out in a regular way the elementary circuits (here just one elementary circuit). These nice diagrams solve the word problem for the surface groups.

6.2.2 Non-orientable surfaces

The non-orientable surface groups are defined by $(A_1, \ldots, A_p \,;\, A_1^2 \cdots A_p^2 = 1)$. There are two cases: p odd or even. We first give the two sets R_3 and R_4:

$$\left\{\begin{array}{rcl}
B^{-2}A^{-1} & \to & CCA \\
A^{-1}C^{-2} & \to & ABB \\
C^{-2}B^{-1} & \to & AAB \\
B^{-1}A^{-2} & \to & BCC \\
A^{-2}C^{-1} & \to & BBC \\
C^{-1}B^{-2} & \to & CAA \\
CCAA & \to & B^{-2} \\
BBCC & \to & A^{-2} \\
CAAB & \to & C^{-1}B^{-1} \\
AABB & \to & C^{-2} \\
ABBC & \to & A^{-1}C^{-1} \\
BCCA & \to & B^{-1}A^{-1} \\
A^{-2}CAA & \to & BBCB^{-2} \\
A^{-1}C^{-1}AAB & \to & ABBC^{-1}B^{-1} \\
B^{-1}A^{-1}BBC & \to & BCCA^{-1}C^{-1} \\
B^{-2}ABB & \to & CCAC^{-2} \\
C^{-1}B^{-1}CCA & \to & CAAB^{-1}A^{-1} \\
C^{-2}BCC & \to & AABA^{-2}
\end{array}\right.$$

$$
\left\{
\begin{array}{lcl}
A^{-2}D^{-2} & \rightarrow & BBCC \\
C^{-1}B^{-2}A^{-1} & \rightarrow & CDDA \\
A^{-1}D^{-2}C^{-1} & \rightarrow & ABBC \\
D^{-2}C^{-2} & \rightarrow & AABB \\
B^{-2}A^{-2} & \rightarrow & CCDD \\
B^{-1}A^{-2}D^{-1} & \rightarrow & BCCD \\
D^{-1}C^{-2}B^{-1} & \rightarrow & DAAB \\
C^{-2}B^{-2} & \rightarrow & DDAA \\
CDDAA & \rightarrow & C^{-1}B^{-2} \\
DDAAB & \rightarrow & C^{-2}B^{-1} \\
BCCDD & \rightarrow & B^{-1}A^{-2} \\
DAABB & \rightarrow & D^{-1}C^{-2} \\
AABBC & \rightarrow & D^{-2}C^{-1} \\
ABBCC & \rightarrow & A^{-1}D^{-2} \\
CCDDA & \rightarrow & B^{-2}A^{-1} \\
BBCCD & \rightarrow & A^{-2}D^{-1} \\
B^{-1}A^{-2}DAAB & \rightarrow & BCCDC^{-2}B^{-1} \\
A^{-2}D^{-1}AABB & \rightarrow & BBCCD^{-1}C^{-2} \\
C^{-1}B^{-2}ABBC & \rightarrow & CDDAD^{-2}C^{-1} \\
A^{-1}D^{-2}CDDA & \rightarrow & ABBCB^{-2}A^{-1} \\
B^{-2}A^{-1}BBCC & \rightarrow & CCDDA^{-1}D^{-2} \\
D^{-2}C^{-1}DDAA & \rightarrow & AABBC^{-1}B^{-2} \\
D^{-1}C^{-2}BCCD & \rightarrow & DAABA^{-2}D^{-1} \\
C^{-2}B^{-1}CCDD & \rightarrow & DDAAB^{-1}A^{-2}
\end{array}
\right.
$$

The general complete set of rules R_p depends on the parity of p. Let $n=[p/2]$, $A_i=A_{i+p}$ if $i=1-p, \ldots, 0$ and $A_i=A_{i-p}$ if $i=p+1, \ldots, 2p$. Both cases split in three sets of rules:

$$p=2n+1$$

$$A_k^{-1}A_{k-1}^{-2} \cdots A_{k-n}^{-2} \rightarrow A_k A_{k+1}^2 \cdots A_{k+n}^2$$

$$A_k^{-2}A_{k-1}^{-2} \cdots A_{k-n+1}^{-2}A_{k-n}^{-1} \rightarrow A_{k+1}^2 \cdots A_{k+n}^2 A_{k+n+1}$$

Note that $A_{k-n}=A_{k+n+1}$.

$$A_k^2 \cdots A_{k+n}^2 \rightarrow A_{k-1}^{-2} \cdots A_{k-n}^{-2}$$

$$A_k A_{k+1}^2 \cdots A_{k+n}^2 A_{k+n+1} \rightarrow A_k^{-1} A_{k-1}^{-2} \cdots A_{k-n+1}^{-2} A_{k-n}^{-1}$$

$$A_k^{-2} \ldots A_{k-n+1}^{-2} A_{k-n} A_{k-n+1}^2 \cdots A_k^2 \rightarrow A_{k+1}^2 \cdots A_{k+n}^2 A_{k+n+1} A_{k+n}^{-2} \cdots A_{k+1}^{-2}$$

$$A_k^{-1} A_{k-1}^{-2} \ldots A_{k-n+1}^{-2} A_{k-n}^{-1} A_{k-n+1}^2 \cdots A_k^2 A_{k+1} \rightarrow A_k A_{k+1}^2 \cdots A_{k+n}^2 A_{k+n-1}^{-1} A_{k+n}^{-2} \cdots A_{k+2}^{-2} A_{k+1}^{-1}$$

$$n = 2p$$

$$A_k^{-2} A_{k-1}^{-2} \cdots A_{k-n+1}^{-2} \rightarrow A_{k+1}^2 \cdots A_{k+n}^2$$

$$A_k^{-1} A_{k-1}^{-2} \cdots A_{k-n+1}^{-2} A_{k-n}^{-1} \rightarrow A_k A_{k+1}^2 \cdots A_{k+n-1}^2 A_{k+n}$$

$$A_k^2 \cdots A_{k+n-1}^2 A_{k+n} \rightarrow A_{k-1}^{-2} \cdots A_{k-n+1}^{-2} A_{k-n}^{-1}$$

$$A_k A_{k+1}^2 \cdots A_{k+n}^2 \rightarrow A_k^{-1} A_{k-1}^{-2} \cdots A_{k-n+1}^{-2}$$

$$A_k^{-2} \ldots A_{k-n+2}^{-2} A_{k-n+1} A_{k-n+2}^2 \cdots A_{k+1}^2 \rightarrow A_{k+1}^2 \cdots A_{k+n}^2 A_{k+n+1}^{-1} A_{k+n}^{-2} \cdots A_{k+2}^{-2}$$

$$A_k^{-1} A_{k-1}^{-2} \ldots A_{k-n+1}^{-2} A_{k-n} A_{k-n+1}^2 \cdots A_{k-1}^2 A_k \rightarrow A_k A_{k+1}^2 \cdots A_{k+n-1}^2 A_{k+n} A_{k+n-1}^{-2} \cdots A_{k+1}^{-2} A_k^{-1}$$

For all these rules, k ranges from 1 to p.
Termination: Lex Ordering with $A_1^{-1} > \cdots > A_p^{-1} > A_1 > \cdots > A_p$.
Every set has $6p$ rules. In both cases, the first four rules are symmetrized presentations, while the last two rules arise from critical pairs with the pieces $A_i \in G \cup G^{-1}$. For example, if $p=4$ the two rules

$$A_1 A_1 A_2 A_2 A_3 \rightarrow A_4^{-1} A_4^{-1} A_3^{-1}$$

$$A_3 A_4 A_4 A_1 A_1 \rightarrow A_3^{-1} A_2^{-1} A_2^{-1}$$

superposed on the piece A_1 creates the new rule:

$$A_3^{-1} A_2^{-1} A_2^{-1} A_1 A_2 A_2 A_3 \rightarrow A_3 A_4 A_4 A_1 A_4^{-1} A_4^{-1} A_3^{-1}.$$

The superposition in $A_1 A_1$, which is not a piece, gives a confluent critical pair according to Prop.5.5, the group satisfying **C'**(4).

6.3 COXETER GROUPS

The word problem for Coxeter groups has been proved decidable by J. Tits [Tit69] with an algorithm enumerating the finite set of derivatives of a word under a relation generated by a finite set of rules. This reduction is confluent but not noetherian, more precisely, if $\lambda \rightarrow \rho$ is a rule, then $\rho \rightarrow \lambda$ also. The completion strengthens this relation into a noetherian one. However, this new relation does not handle groups with commuting pairs of generators.

The completion of these group presentations is perhaps the more convincing example of the power of rewriting systems. By elementary combinatorial methods, it proves the solvability of the word problem whereas geometrical methods are usually used [Bou78]. Furthermore, the family is parametrized by $n \times n$ symmetric integer matrices. Despite this high number of parameters, a terse description of complete presentations is found that leads to an efficient word problem algorithm, as for torus groups. However, a drawback is encountered. The partial commutativity of a presentation leads to a failure, this situation is very close to the general completion procedure where specialized algorithms have been presented to handle both commutativity and associativity (cf §2.3).

6.3.1 Definition of Coxeter groups

See [Bou78] for a detailed analysis of these groups. Let I be a finite set of **n** generators. A Coxeter matrix on I is a function M: $I \times I \rightarrow \mathbb{N} \cup \{\infty\}$ such that for all i,j in I, $M(i,i)=1$ and $M(i,j)=M(j,i) \geq 2$ if $i \neq j$. The value $M(i,j)$ will be denoted by m_{ij}. The Coxeter Group $C(M)$ is presented by (I,E) where E is the set of equations $(ij)^{m_{ij}}=1$, $m_{ij} \neq \infty$. As $m_{ii}=1$ implies $i^{-1}=i$, we may represent the elements of $C(M)$ by words from the free monoid I^* on I. If the words w and w' define the same abstract element of $C(M)$ we write $w =_M w'$. Syntactic equality (equality in I^*) is noted as usual $w=w'$.

Throughout this section, $[ij]^k$ will denote the product $ijij \cdots$ of k generators alternatively equal to i and j; α will denote $[ij]^{m_{ij}-1}$. The generators i and j will be denoted by $f(\alpha)$ and $s(\alpha)$ respectively, and $l(\alpha)$ is the last generator of α, equal to i (resp. j) when m_{ij} is even (resp. odd). To α is associated the word $\bar{\alpha}=[ji]^{m_{ij}-1}$. Finally, m_{ij} will be abbreviated in m_α. The same definitions stand for $\beta=[ij]^{m_{ij}}$ and $\gamma=[ij]^{m_{ij}-2}$.

A first solution to the word problem is given by a theorem due to J. Tits (Thm 1, p.93 of [Bou78]). If the generators from I are interpreted by the following linear transformations of a real vector space with basis $e_1, \ldots, e_n$:

$$s_i : e_j \to e_j - 2(\cos\frac{\pi}{m_{ij}})e_i$$

then a word $w = i_1 \cdots i_k$ from I^* represents the unit element in $C(M)$ iff $s_{i_1} \cdots s_{i_k}(\sum_{j=1}^{n} e_j) = \sum_{j=1}^{n} e_j$. As noted by J. Tits [Tit69], this solution is not efficient. A second solution was proposed in [Tit69] based upon a reduction in $L(I)$ defined by the following rules:

$wiiw' \to ww'$, $i \in I$, $w, w' \in I^*$.

$w\beta w' \to w\bar{\beta}w'$, $w, w' \in I^*$.

The confluence of this system is proved via the linear representation of Coxeter groups. As the reduction is not length increasing, the enumeration of the words reduced from a given one w halts, we can check whether or not $w =_M 1$. The Knuth-Bendix completion may be used to improve significantly this algorithm.

6.3.2 Completion of a Coxeter group

We now detail the completion of a Coxeter group defined by a matrix $M = (m_{ij})$, under a given lexicographic ordering. Together with a constant set of rules, the completion generates new rules sharing a common structure described by a single meta-rule. As already observed at the beginning of this chapter, the complete proof of correctness of the set of rules requires three steps:

1) The reduction is noetherian,
2) Every rule is a consequence of the definition of the group,
3) All critical and normal pairs are solved.

Let us prove these three steps for this completion. The first point follows from our choice of the lexicographic ordering. Also, we prove that the meta-rule is a consequence of the definitions. Then, we will check that all critical and normal pairs are solved or generate an instance of the meta-rule.

6.3.2.1 The meta-rule

First, we restrict ourselves to matrices having no entry equal to 2. The completion begins by symmetrizing the given presentation:

Lemma 6.3

Given a Coxeter matrix M on the set I totally ordered by $>$, the completion generates the following two sets of rules:

$$R_I = \{ i^{-1} \to i,\ ii \to 1 \mid i \in I \} \text{ and } S_I = \{ \beta \to \bar{\beta} \mid f(\beta) > s(\beta) \}.$$

Proof. R_I is generated by the defining relations $ii=1$ and their normal pairs for all i in I. We have $ii>1$, the pairs are (i^{-1},i) and $(i^{-1}i^{-1},1)$. Putting $i^{-1}>i$, the rule $i^{-1}\to i$ is generated, under which the second pair is confluent.

S_I is generated by a sequence of normal pairs from the defining rules. If m_{ij} is even, this rule is $\beta\beta\to 1$. It generates $\beta\alpha\to l(\beta)$. The first rule is redundant and deleted as $\beta\beta$ reduces to $l(\beta)l(\beta)$, then to 1, by the new rule and R_I. This sequence of operations loops and reaches a pair $(\beta,\bar{\beta})$. Then, a rule of type S_I is produced. The case m_{ij} odd is similar ▪

From R_I we can restrict ourselves to words in $I^{\bullet}$. These rules are length decreasing, while those from S_I shift generators i and j. Note that a word α is both a left member suffix (resp. prefix) and a right member prefix (resp. suffix) of a rule in S_I when $f(\alpha)<s(\alpha)$ (resp. >), and that both $\alpha\alpha$ and $\alpha\bar{\alpha}$ reduce on 1.

Theorem 6.4

Let M be a Coxeter matrix on the set I totally ordered by >. If $m_{ij}\neq 2$, $i,j\in I$, then the completion procedure generates the set of rules $R_I\cup T_I$, where T_I consists of all rules of the form

$$\alpha_1\cdots\alpha_n l(\overline{\alpha_n})\to s(\alpha_1)\alpha_1\cdots\alpha_n \tag{T}$$

where n is a positive integer, and for all p such that $1<p<n$:

$$f(\alpha_1)>s(\alpha_1),\ s(\alpha_p)>f(\alpha_p),\ f(\alpha_{p+1})\neq l(\alpha_p),\ s(\alpha_{p+1})=l(\overline{\alpha_p}) \tag{C}$$

A R-rule (resp. S, T) means a rule from R_I (resp. S_I, T_I).

Proof. By C, both words $\beta_n=\alpha_n l(\overline{\alpha_n})$ when $n>1$ and $\overline{\beta_1}=s(\alpha_1)\alpha_1$ are right members of a S-rule. Taking $n=1$, we have $S_I\subset T_I$, and from lemma 2 the completion generates R_I and S_I.

The first step is to prove that every instance of the meta-rule is a consequence of the definitions. This is achieved by an induction on n, lemma 2 claims the result for $n=1$. And we have the following identities:

$$\begin{aligned}
&\alpha_1\cdots\alpha_{n-1}\alpha_n l(\overline{\alpha_n})\\
&=_M \alpha_1\cdots\alpha_{n-1}s(\alpha_n)\alpha_n \quad \text{by } \beta_n=_M\overline{\beta_n}\\
&=\alpha_1\cdots\beta_{n-1}\alpha_n \quad \text{by C and } \beta_{n-1}=\alpha_{n-1}l(\overline{\alpha_{n-1}}).\\
&=s(\alpha_1)\alpha_1\cdots\alpha_{n-1}\alpha_n \quad \text{by induction hypothesis.}
\end{aligned}$$

Thus, the T-rules are consequences of the definitions. The main point is to prove that all normal and critical pairs are solved. This is checked by an analysis of all superpositions. The analysis will also detail how the T-rules are produced. We need an assumption on the sequence of superpositions:

Assumption: Any T-rule is first superposed with S-rules, then on other T-rules.

By induction on the number of αs in a T-rule left member, we show that the completion effectively generates these pairs and only these ones. The induction hypothesis states that the completion limited to the critical and normal pairs of $R_n = R_I \cup \{ u \rightarrow v \mid u \rightarrow v \in T_I,\ u$ has at most n αs $\}$ generates R_{n+1}. We distinguish four superpositions:

i) The superpositions with matching a single generator.

ii) The superpositions with R-rules.

iii) The superpositions with matching length greater than one.

iv) The normal pairs.

The following table will help the reader to keep trace of identities between the generators:

	m_α *even*	m_α *odd*
α	iji	$ijij$
$\overline{\alpha}$	jij	$jiji$
β	$ijij$	$ijiji$
$\overline{\beta}$	$jiji$	$jijij$

Case i.1 Let r be a T-rule $\alpha_1 \cdots \beta_n \rightarrow \overline{\beta_1} \cdots \alpha_n$. We superpose this rule on a S-one $\overline{\beta_{n+1}} \rightarrow \beta_{n+1}$, the matching subword is $l(\overline{\alpha_n}) = f(\overline{\beta_{n+1}})$.

We have two subcases, first $l(\alpha_n) \neq s(\overline{\beta_{n+1}})$. This is the only case creating rules. The least common instance reduces on both:

$$\alpha_1 \cdots \alpha_n l(\overline{\alpha_n})\alpha_{n+1} \rightarrow s(\alpha_1)\alpha_1 \cdots \alpha_n \alpha_{n+1} = A \quad \text{by rule } r.$$

$$\rightarrow \alpha_1 \cdots \alpha_n \beta_{n+1} \quad \text{by } \overline{\beta_{n+1}} \rightarrow \beta_{n+1}$$

$$= \alpha_1 \cdots \alpha_n \alpha_{n+1} l(\overline{\alpha_{n+1}}) = B$$

The critical pair (A,B) has the expected structure, the link condition being satisfied between α_n and α_{n+1}. But such a pair must be non-confluent in order to create a rule. We prove its irreducibility under the current set of rules R_n. Both words

$s(\alpha_1)\alpha_1 \cdots \alpha_n$ and $\alpha_{n+1}l(\overline{\alpha_{n+1}})$ are irreducible as right members. Condition C being satisfied, no R-reduction is possible. Thus, any S or T-reduction must reduce a suffix of α_n and a prefix of α_{n+1}, but the concatenation of these words looks like $\cdots ij.ki \cdots$ with $k \neq i,j$. We cannot reduce generator i as, for A, we do not have the last generator $l(\overline{\alpha_{n+1}})$, and for B we must use a T-rule of $n+1$ components by inequalities C, which is impossible by induction hypothesis. The only possibility is a reduction by the S-rule $jk \rightarrow kj$ if it exists. But this is impossible by the assumption $m_{jk} \neq 2$. Reciprocally, if the pair $p=(\alpha_1 \cdots \beta_{n+1}, \overline{\beta_1} \cdots \alpha_{n+1})$ is an instance of the meta-rule, then both rules $\overline{\beta_{n+1}} \rightarrow \beta_n$ and $\alpha_1 \cdots \beta_n \rightarrow \overline{\beta_1} \cdots \alpha_n$ are generated by induction hypothesis. As C holds for p, $f(\overline{\beta_{n+1}})=s(\alpha_{n+1})=l(\overline{\alpha_n})$. The superposition of the two rules with matching $f(\overline{\beta_{n+1}})$ creates the rule p as detailed just above. Thus, at least all T-rules are effectively generated. We now check that no other ones are generated.

In the subcase where $l(\alpha_n)=s(\overline{\beta_{n+1}})$, m_{α_n} is even and $\alpha_n=\alpha_{n+1}$. Thus, $\alpha_n\alpha_{n+1} \xrightarrow[R]{*} 1$:

$$A \xrightarrow[R]{*} s(\alpha_1)\alpha_1 \cdots \alpha_{n-1}$$

$$B \xrightarrow[R]{*} \alpha_1 \cdots \alpha_{n-1}l(\overline{\alpha_{n+1}})$$

$$= \alpha_1 \cdots \alpha_{n-1}l(\overline{\alpha_{n-1}}) \quad \text{by } l(\overline{\alpha_{n+1}})=l(\overline{\alpha_n})=s(\alpha_n) \text{ and C.}$$

$$\xrightarrow[T]{} s(\alpha_1)\alpha_1 \cdots \alpha_{n-1}=A \quad \text{by induction hypothesis.}$$

The critical pair is confluent in this subcase.

Case i.2 The left part of the least common instance comes from a S-rule. With $f(\alpha_0)>s(\alpha_0)$, $f(\alpha_1)>s(\alpha_1)$ and $l(\overline{\alpha_0})=f(\alpha_1)$, we have:

$$\alpha_0\alpha_1 \cdots \alpha_{n-1}\beta_n \rightarrow \overline{\beta_0}\overline{\gamma_1}\alpha_2 \cdots \alpha_{n-1}\beta_n=A \quad \text{by } \beta_0 \rightarrow \overline{\beta_0}.$$

$$\rightarrow \alpha_0 s(\alpha_1)\alpha_1 \cdots \alpha_n \quad \text{by right reduction.}$$

$$= \alpha_0\overline{\alpha_1}l(\alpha_1)\alpha_2 \cdots \alpha_n=B$$

We then have two subcases. First $f(\overline{\alpha_1})=s(\alpha_1) \neq l(\alpha_0)$:

$$B \xrightarrow[T]{} s(\alpha_0)\alpha_0\overline{\alpha_1}\alpha_2 \cdots \alpha_n \quad \text{by induction hypothesis (C holds for } \alpha_0 \text{ and } \overline{\alpha_1}\text{).}$$

$$= \overline{\beta_0}\overline{\gamma_1}l(\overline{\alpha_1})\alpha_2 \cdots \alpha_n$$

$$= \overline{\beta_0}\overline{\gamma_1}\overline{\beta_2}\alpha_3 \cdots \alpha_n \quad \text{by } s(\alpha_2)=l(\overline{\alpha_1}).$$

$$\rightarrow \overline{\beta_0}\overline{\gamma_1}\alpha_2 l(\overline{\alpha_2})\alpha_3 \cdots \alpha_n \quad \text{by } \overline{\beta_2} \rightarrow \beta_2 \text{ as } s(\alpha_2)>f(\alpha_2).$$

$\cdots$

$\rightarrow\beta_0\overline{\gamma_1}\alpha_2\cdots\alpha_n l(\overline{\alpha_n})=A.$

The reduction $l(\overline{\alpha_1})\alpha_2\cdots\alpha_n \overset{*}{\underset{S}{\Rightarrow}} \alpha_2\cdots\alpha_n l(\overline{\alpha_n})$ will be used again and called a shift/reduce sequence.

Second, $f(\overline{\alpha_1})=s(\alpha_1)=l(\alpha_0)$. This equality implies $\alpha_0=\overline{\alpha_1}$. Thus $\alpha_0\overline{\alpha_1}\overset{*}{\underset{R}{\Rightarrow}}1$ and $\overline{\beta_0\gamma_1}\overset{*}{\underset{R}{\Rightarrow}}l(\alpha_1)l(\overline{\alpha_1})$.

$$A \overset{*}{\underset{R}{\Rightarrow}} l(\alpha_1)l(\overline{\alpha_1})\alpha_2\cdots\alpha_n l(\overline{\alpha_n})$$

$$\overset{*}{\underset{S}{\Rightarrow}} l(\alpha_1)\alpha_2\cdots\alpha_n l(\overline{\alpha_n})l(\overline{\alpha_n}) \quad \text{by shift/reduce.}$$

$$\underset{R}{\Rightarrow} l(\alpha_1)\alpha_2\cdots\alpha_n$$

$$B \overset{*}{\underset{R}{\Rightarrow}} l(\alpha_1)\alpha_2\cdots\alpha_n$$

In both cases, the critical pair is confluent.

Case i.3 We superpose two T-rules on one letter, say the former with n, the latter with m components $1<m\leq n$. With $l(\overline{\alpha_n})=f(\alpha'_1)$ and $f(\alpha'_1)>s(\alpha'_1)$, we have:

$$\alpha_1\cdots\alpha_n\alpha'_1\cdots\alpha'_m l(\overline{\alpha_m})$$

$$\rightarrow s(\alpha_1)\alpha_1\cdots\alpha_n\overline{\gamma'_1}\alpha'_2\cdots\beta'_m=A \quad \text{by left reduction.}$$

$$\rightarrow\alpha_1\cdots\alpha_n s(\alpha'_1)\alpha'_1\cdots\alpha'_m \quad \text{by right reduction.}$$

$$=\alpha_1\cdots\alpha_n\overline{\alpha'_1}l(\alpha'_1)\alpha'_2\cdots\alpha'_m=B$$

As for **i.2**, we have a first case when $f(\overline{\alpha'_1})\neq l(\alpha_n)$. Then C holds for α_n and $\overline{\alpha'_1}$:

$$B \underset{T}{\Rightarrow} s(\alpha_1)\alpha_1\cdots\alpha_n\overline{\alpha'_1}\alpha'_2\cdots\alpha'_m \quad \text{by induction hypothesis.}$$

$$= s(\alpha_1)\alpha_1\cdots\alpha_n\overline{\gamma'_1}l(\overline{\alpha'_1})\alpha'_2\cdots\alpha'_m$$

$$\overset{*}{\underset{S}{\Rightarrow}} s(\alpha_1)\alpha_1\cdots\alpha_n\overline{\gamma'_1}\alpha'_2\cdots\alpha'_m l(\alpha'_m)=A \quad \text{by shift/reduce.}$$

When $f(\overline{\alpha'_1})=l(\alpha_n)$, we have $\alpha_n\overline{\alpha'_1}\overset{*}{\underset{R,S}{\Rightarrow}} 1$ and $\alpha_n\overline{\gamma'_1}\overset{*}{\underset{R,S}{\Rightarrow}} l(\gamma'_1)=l(\overline{\alpha'_1})$. Therefore,

$$A \overset{*}{\underset{R,S}{\Rightarrow}} s(\alpha_1)\alpha_1\cdots\alpha_{n-1}\alpha'_2\cdots\alpha'_m l(\overline{\alpha'_m})l(\overline{\alpha'_m}) \quad \text{by shift/reduce.}$$

$$\underset{R}{\Rightarrow} s(\alpha_1)\alpha_1\cdots\alpha_{n-1}\alpha'_2\cdots\alpha'_m$$

$$B \overset{*}{\underset{R,S}{\Rightarrow}} \alpha_1\cdots\alpha_{n-1}l(\alpha'_1)\alpha'_2\cdots\alpha'_m$$

$$=\alpha_1\cdots\alpha_{n-1}l(\overline{\alpha_{n-1}})\alpha'_2\cdots\alpha'_m$$

$$\underset{T}{\Rightarrow} s(\alpha_1)\alpha_1\cdots\alpha_{n-1}\alpha'_2\cdots\alpha'_m \quad \text{by induction hypothesis.}$$

And the critical pair is confluent in both cases.

Case ii.1 The R-rule appears on the right of the least common instance.

$$\alpha_1 \cdots \alpha_n l(\overline{\alpha_n}) l(\overline{\alpha_n}) \rightarrow s(\alpha_1)\alpha_1 \cdots \alpha_n l(\overline{\alpha_n}) \text{ by left reduction.}$$

$$\rightarrow s(\alpha_1)s(\alpha_1)\alpha_1 \cdots \alpha_n \text{ yet by left reduction.}$$

$$\underset{R}{\rightarrow} \alpha_1 \cdots \alpha_n$$

$$\rightarrow \alpha_1 \cdots \alpha_n \text{ by right reduction.}$$

Case ii.2 The R-rule appears on the left hand side.

$$f(\alpha_1)\alpha_1 \cdots \alpha_n l(\overline{\alpha_n}) \rightarrow \overline{\gamma_1}\alpha_2 \cdots \alpha_n l(\overline{\alpha_n}) \text{ by left reduction.}$$

$$\rightarrow f(\alpha_1)s(\alpha_1)\alpha_1 \cdots \alpha_n \text{ by right reduction.}$$

$$= \beta_1 l(\alpha_1)\alpha_2 \cdots \alpha_n$$

$$\underset{S}{f}\ \overline{\beta_1} l(\alpha_1)\alpha_2 \cdots \alpha_n \text{ as } f(\alpha_1) > s(\alpha_1) \text{ by } C$$

$$\underset{R}{\rightarrow} \overline{\alpha_1}\alpha_2 \cdots \alpha_n \text{ as } l(\overline{\beta_1}) = l(\alpha_1)$$

$$= \overline{\gamma_1} l(\overline{\alpha_1})\alpha_2 \cdots \alpha_n$$

$$\overset{*}{\underset{S}{\rightarrow}} \overline{\gamma_1}\alpha_2 \cdots \alpha_n l(\overline{\alpha_n}) \text{ by shift/reduce.}$$

The critical pair is confluent in both cases.

Case iii.1 The two rules are $\alpha_1 \cdots \beta_n \rightarrow \overline{\beta_1} \cdots \alpha_n$ and $\overline{\beta_n} \rightarrow \beta_n$. Let $i = l(\overline{\alpha_n})$, $j = l(\alpha_n)$ and $m = m_{ij}$. If $0 < k < m$ and k even, we have:

$$\alpha_1 \cdots \alpha_n [ij]^k$$

$$\rightarrow s(\alpha_1)\alpha_1 \cdots \alpha_n [ji]^{k-2} j \text{ by left reduction.}$$

$$\overset{*}{\underset{R}{\rightarrow}} s(\alpha_1)\alpha_1 \cdots \alpha_{n-1}[f(\alpha_n)s(\alpha_n)]^{m-k} \text{ by } k-1 \text{ reductions.}$$

$$\rightarrow \alpha_1 \cdots \alpha_{n-1}[f(\alpha_n)s(\alpha_n)]^{k-1}\beta_n \text{ by right reduction.}$$

$$\overset{*}{\underset{R}{\rightarrow}} \alpha_1 \cdots \alpha_{n-1}[s(\alpha_n)f(\alpha_n)]^{m-k+1} \text{ by } k-1 \text{ reductions.}$$

$$= \alpha_1 \cdots \alpha_{n-1} s(\alpha_n)[f(\alpha_n)s(\alpha_n)]^{m-k}$$

$$= \alpha_1 \cdots \alpha_{n-1} l(\overline{\alpha_{n-1}})[f(\alpha_n)s(\alpha_n)]^{m-k}$$

$$\underset{T}{\rightarrow} s(\alpha_1)\alpha_1 \cdots \alpha_{n-1}[f(\alpha_n)s(\alpha_n)]^{m-k} \text{ by induction hypothesis.}$$

If $i=f(\alpha_1)$, $j=s(\alpha_1)$ and $m=m_{ij}$, the S-rule is now $\beta_1 \rightarrow \overline{\beta_1}$. If $0<k<m$ and k even, we have:

$$[ij]^k \alpha_1 \cdots \alpha_n l(\overline{\alpha_n})$$

$$\rightarrow \overline{\beta_1}[l(\alpha_1)l(\overline{\alpha_1})]^{k-1}\alpha_2 \cdots \alpha_n l(\overline{\alpha_n}) \quad \text{by left reduction.}$$

$$\xrightarrow[R]{*} [ji]^{m-k} l(\overline{\alpha_1})\alpha_2 \cdots \alpha_n l(\overline{\alpha_n}) \quad \text{by } k-1 \text{ reductions.}$$

$$\xrightarrow[S]{*} [ji]^{m-k} \alpha_2 \cdots \alpha_n l(\overline{\alpha_n}) l(\overline{\alpha_n}) \quad \text{by shift/reduce.}$$

$$\xrightarrow[R]{} [ji]^{m-k} \alpha_2 \cdots \alpha_n$$

$$\rightarrow [ij]^k s(\alpha_1)\alpha_1 \cdots \alpha_n \quad \text{by right reduction.}$$

$$\xrightarrow[R]{*} [ji]^{m-k} \alpha_2 \cdots \alpha_n \quad \text{by } k \text{ reductions.}$$

We have confluence in both cases.

Case iii.2 We superpose on more than one generator two T-rules with left members $\alpha_1 \cdots \beta_n$ and $\alpha'_1 \cdots \beta'_m$ with $1<m\leq n$. By the first inequality C, the matching word is a proper subword of α_n. Let $i=l(\alpha_n)$, $j=l(\overline{\alpha_n})$ and $m=m_{ij}$. We have $f(\alpha'_1)=s(\alpha_n)$, $s(\alpha'_1)=f(\alpha_n)$ and $l(\alpha'_1)=l(\overline{\alpha_n})$.

$$\alpha_1 \cdots \alpha_n j [ij]^k \alpha'_2 \cdots \beta_m \quad \text{where } 0\leq k<m-2 \text{ and } k \text{ even.}$$

$$\rightarrow s(\alpha_1)\alpha_1 \cdots \alpha_n [ij]^k \alpha'_2 \cdots \beta_m \quad \text{by left reduction.}$$

$$\xrightarrow[R]{*} s(\alpha_1)\alpha_1 \cdots \alpha_{n-1}[f(\alpha_n)s(\alpha_n)]^{m-1-k}\alpha'_2 \cdots \beta'_m \quad \text{by } k \text{ reductions.}$$

$$= s(\alpha_1)\alpha_1 \cdots \alpha_{n-1}[f(\alpha_n)s(\alpha_n)]^{m-k} l(\overline{\alpha'_1})\alpha'_2 \cdots \beta'_m$$

$$\xrightarrow[R,S]{*} s(\alpha_1)\alpha_1 \cdots \alpha_{n-1}[f(\alpha_n)s(\alpha_n)]^{m-k}\alpha'_2 \cdots \alpha'_m \quad \text{by shift/reduce.}$$

$$\rightarrow \alpha_1 \cdots \alpha_{n-1}[f(\alpha_n)s(\alpha_n)]^{k+1} s(\alpha'_1)\alpha'_1 \cdots \alpha'_m \quad \text{by right reduction.}$$

$$\xrightarrow[R]{*} \alpha_1 \cdots \alpha_{n-1}[f(\alpha'_1)s(\alpha'_1)]^{m-k-1}\alpha'_2 \cdots \alpha'_m \quad \text{by } k+1 \text{ reductions.}$$

$$\xrightarrow[T]{} s(\alpha_1)\alpha_1 \cdots \alpha_{n-1}[f(\alpha_n)s(\alpha_n)]^{m-k}\alpha'_2 \cdots \alpha'_m$$

The last step follows from $f(\alpha'_1)=s(\alpha_n)=l(\overline{\alpha_{n-1}})$ and induction hypothesis. It remains to check the normal pairs.

Case iv.1 The rule is $\alpha_1 \cdots \beta_n \rightarrow \overline{\beta_1} \cdots \alpha_n$ A first normal pair $(\alpha_1 \cdots \alpha_n, s(\alpha_1)\alpha_1 \cdots \alpha_n l(\overline{\alpha_n}))$ is trivially confluent. The second one is

$(\overline{\gamma_1}\alpha_2 \cdots \beta_n \, , f(\alpha_1)s(\alpha_1)\alpha_1 \cdots \alpha_n)$. The last word is equal to:

$$\beta_1 l(\alpha_1)\alpha_2 \cdots \alpha_n \, fS \, \overline{\beta_1} l(\alpha_1)\alpha_2 \cdots \alpha_n \quad \text{as } f(\alpha_1)>s(\alpha_1)$$

$$= \overline{\alpha_1} l(\overline{\beta_1}) l(\alpha_1)\alpha_2 \cdots \alpha_n$$

$$\xrightarrow[R]{} \overline{\alpha_1}\alpha_2 \cdots \alpha_n \quad \text{as } l(\overline{\beta_1})=l(\alpha_1)$$

$$\xrightarrow[S]{*} \overline{\gamma_1}\alpha_2 \cdots \alpha_n l(\overline{\alpha_n}) \quad \text{by shift/reduce.}$$

Case iv.2 The last normal pair takes the inverses. The critical pair (A,B) is $(l(\overline{\alpha_n})\alpha_n^{-1} \cdots \alpha_1^{-1}, \alpha_n^{-1} \cdots \alpha_1^{-1}s(\alpha_1))$. Note that if m_α is even then $\alpha^{-1}=\alpha$, otherwise $\alpha^{-1}=\overline{\alpha}$. If m_{α_n} is even, then:

$$A \rightarrow \beta_n \alpha_{n-1}^{-1} \cdots \alpha_1^{-1} \quad \text{by } \overline{\beta_n} \rightarrow \beta_n.$$

$$= \alpha_n l(\overline{\alpha_{n-1}})\alpha_{n-1}^{-1} \cdots \alpha_1^{-1} \quad \text{as } l(\overline{\alpha_n})=s(\alpha_n)=l(\overline{\alpha_{n-1}}).$$

The same reduction sequence applies while m_k is even. Thus there is confluence if all m are even, otherwise:

$$A \xrightarrow[S]{*} \alpha_n \cdots \alpha_{k+1} l(\overline{\alpha_k})\alpha_k^{-1} \cdots \alpha_1^{-1} \quad \text{with } m_{\alpha_k} \text{odd}.$$

But $B=\alpha_n \cdots \alpha_{k+1}\alpha_k^{-1} \cdots \alpha_1^{-1}s(\alpha_1)$ and the normal pair $(l(\overline{\alpha_k})\alpha_k^{-1} \cdots \alpha_1^{-1}, \alpha_k^{-1} \cdots \alpha_1^{-1}s(\alpha_1))$ is resolved by induction hypothesis. If m_{α_n} is odd, then A is irreducible. We have two subcases. If m_{α_1} is even, then

$$B \rightarrow \alpha_n^{-1} \cdots \alpha_2^{-1}s(\alpha_1)\alpha_1 \quad \text{by } \beta_1 \rightarrow \overline{\beta_1}.$$

$$= \alpha_n^{-1} \cdots \alpha_2^{-1}s(\alpha_2)\alpha_1 \quad \text{as } s(\alpha_1)=l(\overline{\alpha_1})=s(\alpha_2).$$

And, starting at α_2^{-1}, this reduction is applied while m_α is odd. Either all remaining ms are odd, in which case there is confluence, or one m_{α_k} is even. Then a T-rule applies whose last component is α_k, the other ones being a sequence of even αs ending with an odd one which surely exists as m_{α_n} is odd. This T-reduction does not halt this new shift/reduce process, it restarts according to the parity of the m_αs. Thus under a loop of S-reductions followed by T-reductions the pair conflues.

If m_{α_1} is odd, the same loop appears starting with a T-reduction. This concludes the proof ▪

Let us see an example (when we give a complete set of rules, we shall omit the first ones from R_I). The group is defined on three generators a, b and c with $m_{ab}=4$, $m_{bc}=5$ and $m_{ca}=6$. We give two complete presentations; the first one is defined by the ordering $b>a>c$:

$$G1 \left\{ \begin{array}{r@{\;\longrightarrow\;}l} baba & abab \\ bcbcb & cbcbc \\ acacac & cacaca \\ bcbcabab & cbcbcaba \\ babcacaca & ababcacac \\ bcbcabacbcbc & cbcbcabacbcb \end{array} \right.$$

Each rule is a T one. The confluence of the system has been mechanically checked on a DPS8-Multics by the system Al-Zebra [LeC84] (as for all examples in this chapter). The complete system has twelve rules. A smaller system is associated with the ordering $a>c>b$:

$$G2 \left\{ \begin{array}{r@{\;\longrightarrow\;}l} abab & baba \\ cbcbc & bcbcb \\ acacac & cacaca \\ acacabcbcb & cacacabcbc \end{array} \right.$$

Thus, the number of rules depends on the ordering. However, this number does not matter if all rules fall under a single parametrized one. The set of rules may be infinite. Here is an example:

$$G3 \left\{ \begin{array}{r@{\;\longrightarrow\;}l} dcd & cdc \\ dbdb & bdbd \\ dada & adad \end{array} \right.$$

The completion of this set creates infinitely many rules of type $dcbdb\,(adabdb)^n d \longrightarrow cdcbdb\,(adabdb)^n,\ n\geq 0$. Thus we have examples of infinite sets of identities defining efficient algorithms. It is worth noting that all T-rules are in Post normal form: they can be written as $Va \longrightarrow bV$, where V is $\alpha_1 \cdots \alpha_n$. Before the study of a reduction algorithm, we examine presentations with commuting pairs of generators.

6.3.2.2 Commuting pairs of generators

First, let M be a Coxeter matrix with infinite m_{ij}s, then the meta-rule T always gives the complete system and theorem 6.3 is valid for M with the convention that no component α exists for i and j. The meta-rule is puzzled only when some entries of M are equal to 2, i.e. $ij=ji$, the generators commute. This case implies that instances of the meta-rule are S-reducible. With the following *complete* system:

$$G4 \left\{ \begin{array}{r} ad \rightarrow da \\ bd \rightarrow db \\ ca \rightarrow ac \\ cd \rightarrow dc \\ cbc \rightarrow bcb \\ cbac \rightarrow bcba \\ cbabcb \rightarrow bcbabc \end{array} \right.$$

The T-rule $cbadc \rightarrow bcbad$ is never created as its members are confluent under the commutativity rules and the T-rule $cbac \rightarrow bcba$. Critical pairs are reducible with S-rules, then with arbitrary T-ones. Moreover, some T-rules are partially reduced by the commutativity laws (cf. in last section B_n complete presentations). As a final drawback, when a Coxeter matrix has infinite coefficients and others equal to 2, new kinds of rules appear:

$$G5 \left\{ \begin{array}{r} ca \rightarrow ac \\ cb \rightarrow bc \\ dad \rightarrow ada \\ dbd \rightarrow bdb \\ dcd \rightarrow cdc \end{array} \right.$$

With $G4$, the completion procedure generates infinitely many rules $dxcd[yx]^n c \rightarrow xdxcd[yx]^n$, $n \geq 0$ where $\{x,y\}=\{a,b\}$. To moderate these negative results, the last section presents some complete systems of Coxeter groups with commuting pairs of generators.

6.3.3 A Reduction Algorithm

The simplest reduction algorithm iterates the search for a left member, and the substitution of right members. This is of little practical interest. We begin with some remarks on rules and overlapping reductions.

Our goal is a reduction algorithm without backward search in a word already scanned and reduced. After a reduction, what are the possible ones overlapping the new right member ? We first restrict our attention to T-rules:

$$W \xrightarrow{*} v\alpha_1 \cdots \alpha_{n-1}\beta_n w$$

$$\rightarrow v\overline{\beta_1}\alpha_2 \cdots \alpha_n w$$

Any reduction of a v suffix also reduces at most the subword $\overline{\beta_1}$ by the condition C. If it reduces a $\overline{\beta_1}$ prefix of at least two letters, then the same condition C between $\overline{\beta_1}$ and the v suffix implies that $v\alpha_1$ is also reducible. In order to avoid backtracking, the reduction strategy is leftmost, keeping the word v irreducible. If the $\overline{\beta_1}$ prefix is a single generator, reductions may occur. With example G1, we have:

$$acaca\ bcbcabab \rightarrow acaca\ \mathbf{c}bcbcaba = acaca\mathbf{c}\ bcbcaba \rightarrow cacacabcbcaba \qquad \text{(E1)}$$

Thus, the algorithm must update a stack of old left redexes. These prefixes are kept with their occurrence number in the reduced part of the input word. Note that all such prefixes must be stored: we can reach a word vpv_1v_2sw where v_2 is the reverse of v_1, vpv_1 is irreducible and ps is a left member. Then by S,R-reductions, we get the word $vpsw$.

On the right side of the right member, inequalities C imply only one possible reduction: a suffix of the last component α_n is a prefix of the first component of the next redex. But before the reduction we had the configuration $v\alpha_1 \cdots \alpha_n l(\overline{\alpha_n})l(\overline{\alpha_n})w'$. Also, to overcome such overlapping reductions, the algorithm will R-reduce on the right of a T-redex before the T-reduction. Therefore, we may restart the search on the last generator of the right member. With the same example G1:

$$bcbcabab\ cacac \rightarrow cbcbcab\mathbf{a}\ cacac = cbcbcab\ \mathbf{a}cacac \rightarrow bcbcabcacaca \qquad \text{(E2)}$$

The two facts that 1) the new generator (c in E1) can only be the last one of a new redex just before the old one and 2) we can skip until the last generator of the right member (a in E2) are the basic points of the reduction algorithm.

Let us now look more closely at the possible R-reductions following a T-one. On the left, we claim that at most one R-reduction can occur. Otherwise, the T-redex would not be the leftmost one as $f(\alpha_1)s(\alpha_1)\alpha_1 \rightarrow \overline{\beta_1}l(\alpha_1)$. With example G1:

$$bc\ bcbcabab \rightarrow bc\ cbcbcaba \xrightarrow{*} cbcaba$$

But $bc\ bcbcabab = bcbcb\ cabab$, and the redex $bcbcabab$ is not the leftmost one. On the right hand of a right member, we may of course have several R-reductions. Here we may observe that the leftmost strategy is also more efficient than the rightmost one:

$$bcbcabab\ a \rightarrow cbcbcaba\ a \rightarrow cbcbcab$$

While $bcbca\ baba \rightarrow bcbca\ abab \rightarrow bcbcbab \rightarrow cbcbcab$

The rightmost strategy induces a shift/reduce process which makes $n+1$ reductions while the leftmost one always produces two reductions, n being the number of components in the leftmost redex.

We must still examine the consequences of R-reductions. On another R-reduction, they are taken into account by a loop deleting the common generators at top and bottom of the unscanned and reduced part of the input word. On a T-rule, a R-reduction may increase the last redex prefix at the top of the redex stack. Thus a sequence of R-reductions must be closed by an update. This update splits into two operations: the removal of prefixes deleted by R-reductions, and possibly a pop operation on the stack, so that its top becomes the current prefix.

These two observations outline the global strategy of an efficient reduction algorithm based on a leftmost strategy. Let us present more formally the main iteration of the algorithm. This loop updates four variables: v the reduced part of the initial word, w the remaining input, r the current T-redex prefix, and s, the stack: list of redex prefixes together with their occurrence in v. Entering the loop, the word is equal to vrw, where $r=\alpha_1 \cdots \alpha_k$, C being satisfied. Building r needs a function ***component*** which recognizes a component in the word w, i.e. $w=\alpha k w'$, where $k \neq l(\alpha)$. Also a boolean function ***link*** returns true if condition C between the last component α_k of r and α is satisfied. Note that this function uses the assumption $m_{ij} \neq 2$. A procedure ***apply*** applies the rule whose left member has just been recognized. This function possibly pops the stack as a R-rule may appear on the left of the redex right member. The update of r, v and the stack s after a R-reduction is performed by a procedure ***update***. Finally, ***plast*** returns the last generator of $\overline{\alpha}$, and, given two generators a,b, ***alpha*** computes α_{ab}. A list is noted [a;b;c], the dot . and the at @ are the list cons and append, ***last*** and ***tail*** are usual list functions.

Main loop of the reduction algorithm

loop { **While** *w=aaw'* **do** { *w:=w'* }; **While** *w=abbw'* **do** { *w:=aw'* }; **If** *a=last(v@r)* **Then** { *w:=tail(w); update(v,r,s)* } **Else** { *(bool, w', c, d):=component(a,b,(tail (tail w)));* **If** *bool* **Then** { *(w starts with a component)* **If** *a>b* **Then** { **If** *c=plast(a,b)* **Then** { *apply([],a,b,(d.w'),(v@r))* }
Else { *v:=v@r; push(s,r); r:=alpha(a b)* }}
Else { **If** *r=[]* **Then** { *v:=v@alpha(a,b)* }
Else { **If** *c=plast(a,b) and link(a,b,r)* **Then** { *apply(r,a,b,(d.w'),v)* }
Else { *r:=r@alpha(a,b)* }}}}
Else { *v:=v@[a;b]; w:=w'; push(s,r); r:=[]* }}} ▪

6.3.4 Some examples

In all examples, the set R_I is omitted. We consider finite Coxeter groups first described by H.S.M Coxeter [Cox35]. The notations are taken from [Bou78]. The completion of a finite group always halts (cf. §3.2). The finite Coxeter groups whose matrix entries are equal to 1,2,3,4 or 6 are called crystallographic groups. However we failed to complete the following crystallographic groups: E_n, n=6,7,8 and the family D_n.

Groups H_4, H_3 and $I_2(n)$.

$$H_4 \left\{ \begin{array}{r} dcd \to cdc \\ cbc \to bcb \\ cbac \to bcba \\ babab \to ababa \\ cbabcb \to bcbabc \\ cbabacbaba \to bcbabacbab \end{array} \right. \qquad H_3 \left\{ \begin{array}{r} cbc \to bcb \\ cbac \to bcba \\ babab \to ababa \\ cbabcb \to bcbabc \\ cbabacbaba \to bcbabacbab \end{array} \right.$$

These two groups are not crystallographic, nor are the dihedral groups $I_2(n)$, $n>4$, except $I_2(6)$. These groups are generated by two plane reflections through lines whose angle is $\frac{2\pi}{n}$. Their complete presentation is the simpler one. All critical pairs are solved by symmetrization. The complete set is $R_I \cup S_I$. The remaining finite groups are crystallographic.

Groups F_4, E_6, E_7 and E_8.

$$F_4 \left\{ \begin{array}{r} bab \to aba \\ dcd \to cdc \\ cbcb \to bcbc \\ cbacba \to bcbacb \end{array} \right.$$

The Knuth-Bendix procedure failed to complete the three groups E_n, n=6,7,8 while generating hundred rules such as, for E_6:

$$\begin{array}{rcl} fcbadcbedfcbadc & \to & cfcbadcbedfcbad \\ fcbadcbedcfcbdcf & \to & cfcbadcbedcfcbdc \end{array}$$

As these groups possess many matrix entries equal to 2, we do not have an efficient reduction algorithm for these three groups.

Groups A_n, B_{n+1} and D_{n+3}, n≥1.

These three families define the infinite families of crystallographic groups. The first family is the family of symmetric groups. Their complete presentation may be found in §6.5. Despite commuting pairs of generators, note that we have only T-rules. The groups B_n also possess a fair complete presentation. Their Coxeter matrix is:

$$\begin{bmatrix} 1 & 3 & & & & \\ 3 & 1 & \ddots & & \text{2} & \\ & \ddots & \ddots & \ddots & & \\ & & \ddots & 1 & 3 & 2 \\ & \text{2} & & 3 & 1 & 4 \\ & & & 2 & 4 & 1 \end{bmatrix}$$

The complete system includes the rules R_I, the rules of commutativity and the following T-rules for B_n:

$$\begin{cases} a_i a_{i-1} \cdots a_{i-k} a_i \to a_{i-1} a_i a_{i-1} \cdots a_{i-k} & n > i > k > 0 \\ (a_n a_{n-1} \cdots a_{n-k})^2 \to a_{n-1} a_n a_{n-1} \cdots a_{n-k} a_n a_{n-1} \cdots a_{n-k+1} & n > k > 0 \end{cases}$$

The last family D_n, however, does not possess an easily described complete system. Their Coxeter matrix is the previous one where the last row and column are replaced by [2 2 ... 2 3 2 1]. Here are the T-rules except the commutative ones for D_4:

$$\begin{cases} dbd \to bdb \\ cbc \to bcb \\ bab \to aba \\ cbac \to bcba \\ dbcb \to cdbc \\ dbad \to bdba \\ dbcd \to bdbc \\ dbacd \to bdbac \\ dbacba \to cdbacb \\ dbacbdb \to bdbacbd \end{cases}$$

In [Gil79], R.H. Gilman proposes a procedure which is basically the completion procedure described here. In particular, his lemma 1.(3) defines the superposition between rules. The main difference with the present completion lies in the computation of normal pairs giving a completion procedure well suited for groups.

6.4 POLYHEDRAL GROUPS

The polyhedral group (l,m,n) is defined by the presentation $(A,B,C\ ;\ ABC,A^l,B^m,C^n)$ [Cox72]. We present in this section complete systems for the following generalization:

$$(p_1,\ldots,p_n)=(A_1,\ldots,A_n;A_1^{p_1},\ldots,A_n^{p_n},A_1\cdots A_n)\quad n>2.$$

Observe that these groups are subgroups of Coxeter groups (the rotation subgroup). As for Coxeter groups, the general complete system requires $p_i>2$, $i=1,\ldots,n$. Let us first examine the case $n=3$. This presentation is redundant; one of the generators, say C, may be eliminated. Then we can see on the new presentation $(A,B;A^l,B^m,(AB)^n)$ that when l,m and n are greater than 3 the group is a small cancellation one, as the only pieces are the generators and their inverses. Thus its word problem is solvable. The groups are infinite when $\frac{1}{l}+\frac{1}{m}+\frac{1}{n}\leq 1$. Thus, the only finite ones are $(2,2,n)$, $(2,3,3)$, $(2,3,4)$, $(2,3,5)$. The first one is the group of rotations and reflections of a regular p-gon of the euclidean plane. Its complete system is quite trivial:

$$(2,2,n)\left\{\begin{array}{lcl} A^{-1} & \rightarrow & A \\ B^{-1} & \rightarrow & B \\ AA & \rightarrow & 1 \\ BB & \rightarrow & 1 \end{array}\right.$$

and, if $n=2p+1$ $(AB)^pA\rightarrow(BA)^pB$

if $n=2p$ $(AB)^p\rightarrow(BA)^p$

The remaining finite groups are the rotation groups of the five regular polyhedrons of the three dimensional space:

Tetraedron:

$$\left\{\begin{array}{lcl} C^{-1} & \rightarrow & CC \\ B^{-1} & \rightarrow & BB \\ BBB & \rightarrow & 1 \\ CCC & \rightarrow & 1 \\ CBC & \rightarrow & BB \\ BCB & \rightarrow & CC \\ CCBB & \rightarrow & BC \\ BBCC & \rightarrow & CB \\ CBBC & \rightarrow & BCCB \end{array}\right.$$

Termination: KB Ordering with $\pi(B^{-1})=3$, $\pi(C^{-1})=3$, $\pi(B)=\pi(C)=1$ and $C>B$ for the last equation.

We also have the following locally confluent system, whose termination is not obvious:

$$\left\{\begin{array}{rcl} B^{-1} & \to & CBC \\ C^{-1} & \to & CC \\ BB & \to & CBC \\ CCC & \to & 1 \\ BCB & \to & CC \\ BCCBC & \to & CCB \\ CCBCC & \to & BCCB \\ CBCCB & \to & BCC \end{array}\right.$$

Cube (or its dual **Octaedron**):

$$\left\{\begin{array}{rcl} A^{-1} & \to & A \\ C^{-1} & \to & ACACA \\ AA & \to & 1 \\ CCC & \to & ACACA \\ CACAC & \to & A \\ CACCAC & \to & ACCA \\ CCACCA & \to & ACCACC \end{array}\right.$$

Termination: KB Ordering with $\pi(C^{-1})=5, \pi(A^{-1})=\pi(C)=\pi(A)=1$, and $C^{-1}>A^{-1}>C>A$.

Icosaedron (or **Dodecaedron**):

$$\left\{\begin{array}{rcl} A^{-1} & \to & A \\ B^{-1} & \to & BB \\ BBB & \to & 1 \\ AA & \to & 1 \\ BABABAB & \to & ABBA \\ BBABB & \to & ABABABA \\ BABABBABAB & \to & ABBABABBA \\ BABBABABBABA & \to & ABABBABABBAB \end{array}\right.$$

Termination: KB Ordering with $\pi(B^{-1})=6, \pi(B)=3, \pi(A^{-1})=\pi(A)=1$, and $A^{-1}>B^{-1}>A>B$.

All the remaining groups are infinite. We may suppose that $l \le m \le n$, because of the symmetry of the presentation (all the groups (l,m,n) and (p,q,r) with $\{l,m,n\}=\{p,q,r\}$ are isomorphic). There are two distinct cases: either 2 is the power of some generator or not. We first give the general system. When no parameter equals 2, this set of rules is divided into three sets of six rules. The rules are given in a parametrized form according to the parity of the parameters. The simpler case occurs when all the parameters are odd: $l=2p+1, m=2q+1, n=2r+1$. The remaining cases have slight modifications of the exponents. The first set is

$$\left\{\begin{array}{rcl} AB & \rightarrow & C^{-1} \\ BC & \rightarrow & A^{-1} \\ CA & \rightarrow & B^{-1} \\ A^{-1}C^{-1} & \rightarrow & B \\ C^{-1}B^{-1} & \rightarrow & A \\ B^{-1}A^{-1} & \rightarrow & C \end{array}\right.$$

These six rules are the symmetrized set of the defining relator ABC. Then we have the six symmetrized rules of the three equations defining the order of the generators.

$$\left\{\begin{array}{rcl} A^{p+1} & \rightarrow & A^{-p} \\ A^{-(p+1)} & \rightarrow & A^{p} \\ B^{q+1} & \rightarrow & B^{-p} \\ B^{-(q+1)} & \rightarrow & B^{p} \\ C^{r+1} & \rightarrow & C^{-r} \\ C^{-(r+1)} & \rightarrow & C^{r} \end{array}\right.$$

Then, the last six rules are the following ones:

$$\left\{\begin{array}{rcl} A^{-1}C^{r} & \rightarrow & BC^{-r} \\ A^{-p}B & \rightarrow & A^{p}C^{-1} \\ B^{-1}A^{p} & \rightarrow & CA^{-p} \\ B^{-q}C & \rightarrow & B^{q}A^{-1} \\ C^{-1}B^{q} & \rightarrow & AB^{-q} \\ C^{-r}A & \rightarrow & C^{r}B^{-1} \end{array}\right.$$

Termination: Lex Ordering with $C^{-1}>B^{-1}>A^{-1}>C>B>A$.

Three cases remain, when one, two or three generators have an even order. The set (1) of rewrite rules remains the same. The set (2) is modified when the order of a generator becomes even. If A has order $2p$, then the two corresponding rules are:

$$\left\{\begin{array}{rcl} A^{p+1} & \rightarrow & A^{-(p-1)} \\ A^{-p} & \rightarrow & A^{p} \end{array}\right.$$

Then, in the last set of rules (3), we now have two modifications:

$$\begin{cases} A^{-(p-1)}B \rightarrow A^p C^{-1} \\ B^{-1}A^p \rightarrow CA^{-(p-1)} \end{cases}$$

These modifications occur for every generator, whatever the order of the parity shift. For example, we give four complete sets (7,7,7), (7,8,9), (7,8,8) and (8,8,8) on next page.

For heuristic purposes, it is of interest to detail the discovery of these complete sets. The right way is to test testing a great number of presentations with random parameters (not too big, however!), under various orderings. The analysis of the results outlines the fair ordering (the one for the odd case), but only the case of odd generator orders was satisfactory, yielding the previous complete set. The other cases computed about thirty rules. A detailed analysis proved that in all the unexpected rules some constant subword from the set (3) appeared, due to an inversion of the rules (3) associated to an even generator. Restoring the right way gave us the previous sets.

But the termination is now a hard problem. At least one, and at most three rules in the even case are length increasing, and no classical ordering proves the noetherianity (for the KB ordering, searching for a solution is to solve the simplex algorithm for eighteen linear inequalities with seven unknowns and three parameters, seven pages of inequations to prove that no solution exists at all!). From hand-checked examples, we conjecture that the reductions are noetherian.

The irreducible forms of (l,m,n) are described by the finite automaton of Figure 6.3 (recall that the set of normal forms is regular), with the following conventions:

- A state labelled A (resp. B,C) recognizes the subwords A^i, $i=1, \ldots, [l/2]$.
  A state labelled a (resp. b,c) recognizes the subwords A^{-i}, $i=1, \ldots, [(l-1)/2]$.
- Simple arrows allow all transitions, whatever the subword recognized by the initial state of the arrow. Double arrows allow all transitions except the one whose initial state has recognized the maximal length subword (rules with left members $B^{-3}C$). Triple arrows allow all transitions except the one whose final state recognizes the maximal subword (rules with left members $B^{-1}A^3$).

$$
\left\{
\begin{array}{rcl}
AB & \rightarrow & C^{-1} \\
BC & \rightarrow & A^{-1} \\
CA & \rightarrow & B^{-1} \\
A^{-1}C^{-1} & \rightarrow & B \\
C^{-1}B^{-1} & \rightarrow & A \\
B^{-1}A^{-1} & \rightarrow & C \\
A^{4} & \rightarrow & A^{-3} \\
A^{-4} & \rightarrow & A^{3} \\
B^{4} & \rightarrow & B^{-3} \\
B^{-4} & \rightarrow & B^{3} \\
C^{4} & \rightarrow & C^{-3} \\
C^{-4} & \rightarrow & C^{3} \\
B^{-1}A^{3} & \rightarrow & CA^{-3} \\
A^{-3}B & \rightarrow & A^{3}C^{-1} \\
C^{-1}B^{3} & \rightarrow & AB^{-3} \\
B^{-3}C & \rightarrow & B^{3}A^{-1} \\
A^{-1}C^{3} & \rightarrow & BC^{-3} \\
C^{-3}A & \rightarrow & C^{3}B^{-1}
\end{array}
\right.
$$

$$
\left\{
\begin{array}{rcl}
AB & \rightarrow & C^{-1} \\
BC & \rightarrow & A^{-1} \\
CA & \rightarrow & B^{-1} \\
A^{-1}C^{-1} & \rightarrow & B \\
C^{-1}B^{-1} & \rightarrow & A \\
B^{-1}A^{-1} & \rightarrow & C \\
A^{4} & \rightarrow & A^{-3} \\
A^{-4} & \rightarrow & A^{3} \\
B^{5} & \rightarrow & B^{-3} \\
B^{-4} & \rightarrow & B^{4} \\
C^{5} & \rightarrow & C^{-4} \\
C^{-5} & \rightarrow & C^{4} \\
B^{-1}A^{3} & \rightarrow & CA^{-3} \\
A^{-3}B & \rightarrow & A^{3}C^{-1} \\
C^{-1}B^{4} & \rightarrow & AB^{-3} \\
B^{-3}C & \rightarrow & B^{4}A^{-1} \\
A^{-1}C^{4} & \rightarrow & BC^{-4} \\
C^{-4}A & \rightarrow & C^{4}B^{-1}
\end{array}
\right.
$$

$$
\left\{
\begin{array}{rcl}
AB & \rightarrow & C^{-1} \\
BC & \rightarrow & A^{-1} \\
CA & \rightarrow & B^{-1} \\
A^{-1}C^{-1} & \rightarrow & B \\
C^{-1}B^{-1} & \rightarrow & A \\
B^{-1}A^{-1} & \rightarrow & C \\
A^{4} & \rightarrow & A^{-3} \\
A^{-4} & \rightarrow & A^{3} \\
B^{5} & \rightarrow & B^{-3} \\
B^{-4} & \rightarrow & B^{4} \\
C^{5} & \rightarrow & C^{-3} \\
C^{-4} & \rightarrow & C^{4} \\
B^{-1}A^{3} & \rightarrow & CA^{-3} \\
A^{-3}B & \rightarrow & A^{3}C^{-1} \\
C^{-1}B^{4} & \rightarrow & AB^{-3} \\
B^{-3}C & \rightarrow & B^{4}A^{-1} \\
A^{-1}C^{4} & \rightarrow & BC^{-3} \\
C^{-3}A & \rightarrow & C^{4}B^{-1}
\end{array}
\right.
$$

$$
\left\{
\begin{array}{rcl}
AB & \rightarrow & C^{-1} \\
BC & \rightarrow & A^{-1} \\
CA & \rightarrow & B^{-1} \\
A^{-1}C^{-1} & \rightarrow & B \\
C^{-1}B^{-1} & \rightarrow & A \\
B^{-1}A^{-1} & \rightarrow & C \\
A^{5} & \rightarrow & A^{-3} \\
A^{-4} & \rightarrow & A^{4} \\
B^{5} & \rightarrow & B^{-3} \\
B^{-4} & \rightarrow & B^{4} \\
C^{5} & \rightarrow & C^{-3} \\
C^{-4} & \rightarrow & C^{4} \\
B^{-1}A^{4} & \rightarrow & CA^{-3} \\
A^{-3}B & \rightarrow & A^{4}C^{-1} \\
C^{-1}B^{4} & \rightarrow & AB^{-3} \\
B^{-3}C & \rightarrow & B^{4}A^{-1} \\
A^{-1}C^{4} & \rightarrow & BC^{-3} \\
C^{-3}A & \rightarrow & C^{4}B^{-1}
\end{array}
\right.
$$

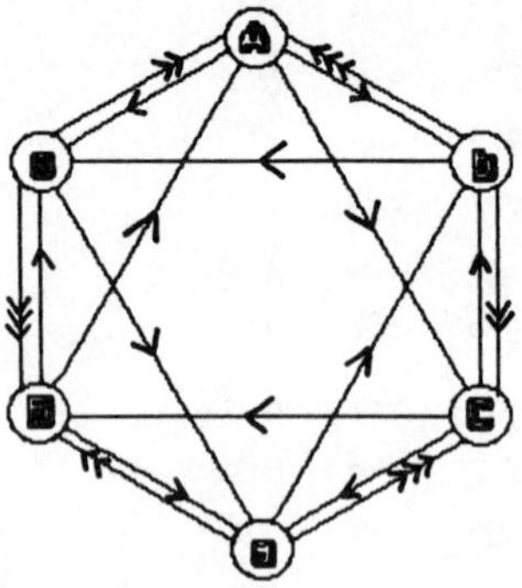

Fig. 6.3

We now describe the complete systems of "polyhedral" groups defined by at least four generators. We have two cases according to the parity of the number of generators. As for Coxeter groups, we restrict ourselves to generators of period greater than 2.

6.4.1 Polyhedral groups with odd number of generators

Let $G=\{A_1,\ldots,A_{2n+1}\}$ be the linearly ordered set of generators. Let $\alpha_1 \cdots \alpha_{n+1}$ be any subword of length $n+1$ in the word $W_G=A_1 \cdots A_{2n+1}A_1 \cdots A_n$. The word $\alpha_{n+2} \cdots \alpha_{2n+1}$ denotes its *complement*: suffix of length n, or prefix of length n if such a suffix does not exist. We give the complete presentation for odd exponents. The complete system for $(2p_1+1,\ldots,2p_{2n+1}+1)$, $p_i>0$, is therefore:

$$\left\{\begin{array}{rcl}
\alpha_1 \cdots \alpha_{n+1} & \rightarrow & (\alpha_{n+2} \cdots \alpha_{2n+1})^{-1} \\
(\alpha_1 \cdots \alpha_{n+1})^{-1} & \rightarrow & \alpha_{n+2} \cdots \alpha_{2n+1} \\
\alpha^{p_\alpha+1} & \rightarrow & \alpha^{-p_\alpha} \\
\alpha^{-(p_\alpha+1)} & \rightarrow & \alpha^{p_\alpha} \\
\alpha_{n+1}^{-1} \cdots \alpha_2^{-1}\alpha_1^{p_{\alpha_1}} & \rightarrow & \alpha_{n+2} \cdots \alpha_{2n+1}\alpha_1^{-p_{\alpha_1}} \\
\alpha^{-p_{\alpha_1}}\alpha_2 \cdots \alpha_{n+1} & \rightarrow & \alpha_1^{p_{\alpha_1}}(\alpha_{n+2} \cdots \alpha_{2n+1})^{-1}
\end{array}\right.$$

For complete presentations with generators α of even order $2p_\alpha$, the third and fourth rules become $\alpha^{p_\alpha+1} \rightarrow \alpha^{-(p_\alpha-1)}$ and $\alpha^{-p_\alpha} \rightarrow \alpha^{p_\alpha}$ respectively. And the other pairs of exponents $(p_\alpha,-p_\alpha)$ become $(p_\alpha,-(p_\alpha-1)$. Here is, for example, the rules for (6,5,5,5,5); upper-case letters denote the generators, lower-case ones their inverses:

$ABC \to ed$ $BCD \to ae$ $CDE \to ba$ $DEA \to cb$ $EAB \to dc$

$aed \to BC$ $bae \to CD$ $cba \to DE$ $dcb \to EA$ $edc \to AB$

$AAAA \to aa$ $BBB \to bb$ $CCC \to cc$ $DDD \to dd$ $EEE \to ee$

$aaa \to AAA$ $bbb \to BB$ $ccc \to CC$ $ddd \to DD$ $eee \to EE$

$aeDD \to BCdd$ $baEE \to CDee$ $cbAAA \to DEaa$ $dcBB \to EAbb$ $edCC \to ABcc$

$aaBC \to AAAed$ $bbCD \to BBae$ $ccDE \to CCba$ $ddEA \to DDcb$ $eeAB \to EEdc$

The number of rules is $6|G|$. As for surface groups, the sets of rules are simply described geometrically. We display the rules in the Cayley graph, starting at a given vertex. We present (5,5,5). For the other groups with odd number $2n+1$ of generators the number of polygons around the central vertex is $4n+2$. The grey regions denote the defining relation $ABC=1$. The Cayley graphs are planar, so that they are oriented according to the displayed arrow. The size of the other polygons depends on the order of the generator. Bold lines denote forbidden edges for paths in normal form.

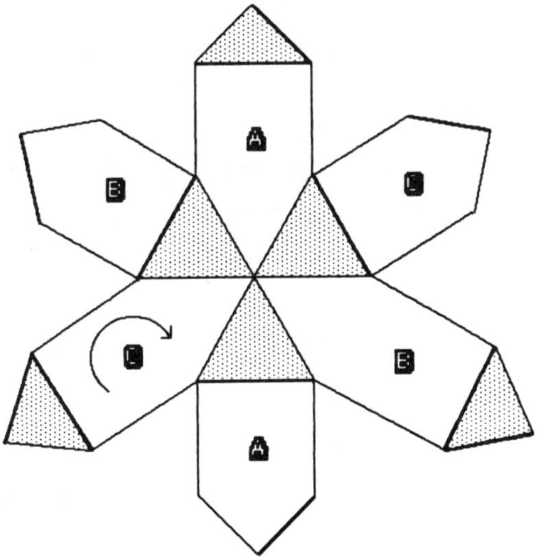

Fig. 6.4, (5,5,5).

6.4.2 Polyhedral groups with even number of generators

The number of rules is now $10|G|$, instead of $6|G|$. Let $2n$ be the number of generators and put $W_G = A_1 \cdots A_{2n} A_1 \cdots A_n$. The words $\alpha_1 \cdots \alpha_{n+1}$ have the same definition as above. We also need the words $\alpha_1 \cdots \alpha_n$ together with their complements $\alpha_{n+1} \cdots \alpha_{2n}$. We give the complete system for generators of odd exponent. The complete system for $(2p_1+1, \ldots, 2p_{2n}+1)$, $p_i > 0$, is:

$$
\begin{cases}
\alpha_1 \cdots \alpha_{n+1} \to (\alpha_{n+2} \cdots \alpha_{2n})^{-1} \\
(\alpha_1 \cdots \alpha_n)^{-1} \to \alpha_{n+1} \cdots \alpha_{2n} \\
\alpha^{p_\alpha+1} \to \alpha^{-p_\alpha} \\
\alpha^{-(p_\alpha+1)} \to \alpha^{p_\alpha} \\
\alpha_1 \cdots \alpha_n \alpha_{n+1}^{-p_{\alpha_{n+1}}} \to (\alpha_{n+2} \cdots \alpha_{2n})^{-1} \alpha_{n+1}^{p_{\alpha_{n+1}}} \\
\alpha_1^{-p_{\alpha_1}} \alpha_2 \cdots \alpha_{n+1} \to \alpha_1^{p_{\alpha_1}} (\alpha_{n+2} \cdots \alpha_{2n})^{-1} \\
\alpha_1^{-p_{\alpha_1}} \alpha_2 \cdots \alpha_n \alpha_{n+1}^{-p_{\alpha_{n+1}}} \to \alpha_1^{p_{\alpha_1}} (\alpha_{n+2} \cdots \alpha_{2n})^{-1} \alpha_{n+1}^{p_{\alpha_{n+1}}} \\
(\alpha_{n+2} \cdots \alpha_{2n})^{-1} \alpha_{n+1}^{p_{\alpha_{n+1}}} (\alpha_2 \cdots \alpha_n)^{-1} \alpha_1^{p_{\alpha_1}} \to \alpha_1 \cdots \alpha_n \alpha_{n+1}^{-(p_{\alpha_{n+1}}-1)} \alpha_{n+2} \cdots \alpha_{2n} \alpha_1^{-p_{\alpha_1}} \\
\alpha_1 \cdots \alpha_n \alpha_{n+1}^{-(p_{\alpha_{n+1}}-1)} \alpha_{n+2} \cdots \alpha_{2n} \alpha_1 \to (\alpha_{n+2} \cdots \alpha_{2n})^{-1} \alpha_{n+1}^{p_{\alpha_{n+1}}} (\alpha_2 \cdots \alpha_n)^{-1} \\
\alpha_1^{-p_{\alpha_1}} \alpha_2 \cdots \alpha_n \alpha_{n+1}^{-(p_{\alpha_{n+1}}-1)} \alpha_{n+2} \cdots \alpha_{2n} \alpha_1 \to \alpha_1^{p_{\alpha_1}} (\alpha_{n+2} \cdots \alpha_{2n})^{-1} \alpha_{n+1}^{p_{\alpha_{n+1}}} (\alpha_2 \cdots \alpha_n)^{-1}
\end{cases}
$$

For generators with an even exponent, the observations of the previous section remain valid, together with the convention that $-(p_\alpha - 1)$ becomes $-(p_\alpha - 2)$ for a generator α of exponent $2p_\alpha$. We give for example (6,5,5,5):

$ABC \to d$	$BCD \to a$	$CDA \to b$	$DAB \to c$
$ad \to BC$	$ba \to CD$	$cb \to DA$	$dc \to AB$
$AAAA \to aa$	$BBB \to bb$	$CCC \to cc$	$DDD \to dd$
$aaa \to AAA$	$bbb \to BB$	$ccc \to CC$	$ddd \to DD$
$ABcc \to dCC$	$BCdd \to aDD$	$CDaa \to bAAA$	$DAbb \to cBB$
$aaBcc \to AAAdCC$	$bbCdd \to BBaDD$	$ccDaa \to CCbAAA$	$ddAbb \to DDcBB$
$ABcDA \to dCCb$	$BCdAB \to aDDc$	$CDaBC \to bAAAd$	$DAbCD \to cBBA$
$aDDcBB \to BCdAbb$	$bAAAdCC \to CDaBcc$		
$cBBaDD \to DAbCdd$	$dCCbAAA \to ABcDaa$		
$aaBcDA \to AAAdCCb$	$bbCdAB \to BBaDDc$		
$ccDaBC \to CCbAAAd$	$ddAbCD \to DDcBBa$		

As in the previous section, we give a geometrical interpretation of the rules, here for

(5,5,5,5) :

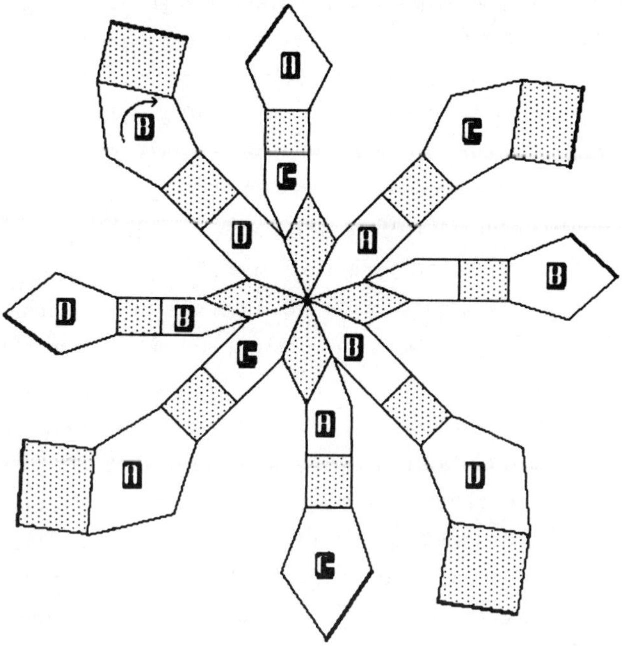

Fig. 6.5, (5,5,5,5).

When the number of generators increase, so does the number of branches around the central vertex. And the number of edges in polygons varies as the exponent of generators. Observe that the initial cycles surrounding the central vertex give two rules, the others only one. These figures give a concise construction of Cayley graphs. The critical pairs are computed by superposition on a ***single*** generator, as for Coxeter groups. The completion procedure stops as the remaining superposition creates a subgraph yet appearing somewhere else in the graph. Finally, remember that for presentations including generators with even period, we did not prove the termination of the system due to length-increasing rules, such as $aaBcDA \longrightarrow AAAdCCb$ in (6,5,5,5).

6.5 SYMMETRIC GROUPS

The symmetric group S_n of order n is the group of all permutations of a sequence of n distinct objects, or equivalently the set of all bijections of a set with n elements. They may be represented by a sequence of n integers $(i_j)_{j=1...n}$, $i_j \le n$ and $i_j \ne i_k$ for $j \ne k$, this permutation maps j on i_j. There are $n!$ such permutations, the ***transpositions*** are defined by $i_j = j$ except for $1 \le k < l \le n$, $i_k = l$ and $i_l = k$. A transposition exchanges two elements, fixing the other ones. It is a well-known fact, currently used by sorting algorithms, that the transpositions generate the symmetric group. Even the *adjacent* transpositions such that $l = k+1$ generate the whole group. A terse notation for transpositions is $(l\ k)$. More generally a cycle is a sequence of length $m \le n$ $(l_1 l_2 ... l_m)$ that represents the permutation $\sigma(l_k) = l_{k+1}$, and $\sigma(i) = i$ if i does not appear in the sequence. Any permutation decomposes into cycles. The transpositions are cycles of length 2; the smaller generating set for S_n is a transposition with a cycle of length n.

Given a finite group $\mathbf{G}$ with n elements, the map $a \in \mathbf{G} \longrightarrow T_a$, where T_a is the permutation of $\mathbf{G}$ $T_a(x) = a.x$ shows that $\mathbf{G}$ is a subgroup of $\mathbf{S}_n$. This result, known as Cayley theorem, shows the prominent part of symmetric groups in finite group theory. See for example the monograph on Group Theoretic Algorithms by C.H. Hoffmann [Hof82] where algorithms are expressed in terms of ***permutation*** groups, or subgroups of $\mathbf{S}_n$. We will give two complete presentations of the symmetric group. The first one is based on a presentation by adjacent transpositions

6.5.1 Presentation with adjacent transpositions

The presentation $\mathbf{S}_n$ by adjacent transpositions is the following one:

$$\begin{cases} R_i = (i\ i{+}1) & i=1,...,n-1 \\ R_i^2 = 1 & i=1,...,n-1 \\ R_i R_j = R_j R_i & i \le j-2 \\ (R_i R_{i+1})^3 = 1 & i \le n-2 \end{cases}$$

The completion gives the n^2-2n+2 rules:

$$\begin{cases} R_i^{-1} \rightarrow R_i & i=1,...,n \\ R_i^2 \rightarrow 1 & i=1,...,n \\ R_i R_j \rightarrow R_j R_i & j \le i-2 \\ R_i R_{i-1} \cdots R_j R_i \rightarrow R_{i-1} R_i R_{i-1} \cdots R_j & j<i \end{cases}$$

Let $1 = R_0$ in S_n then for each rule the integer made by the concatenation of the left members generators indices is greater than the right member one, thus the system is noetherian.

The whole set of rules for S_5 is:

$$\left\{\begin{array}{rcl}
R_1^{-1} & \rightarrow & R_1 \\
R_2^{-1} & \rightarrow & R_2 \\
R_3^{-1} & \rightarrow & R_3 \\
R_4^{-1} & \rightarrow & R_4 \\
R_1R_1 & \rightarrow & 1 \\
R_2R_2 & \rightarrow & 1 \\
R_3R_3 & \rightarrow & 1 \\
R_4R_4 & \rightarrow & 1 \\
R_3R_1 & \rightarrow & R_1R_3 \\
R_4R_1 & \rightarrow & R_1R_4 \\
R_2R_4 & \rightarrow & R_2R_4 \\
R_2R_1R_2 & \rightarrow & R_1R_2R_1 \\
R_3R_2R_3 & \rightarrow & R_2R_3R_2 \\
R_4R_3R_4 & \rightarrow & R_3R_4R_3 \\
R_3R_2R_1R_3 & \rightarrow & R_2R_3R_2R_1 \\
R_4R_3R_2R_4 & \rightarrow & R_3R_4R_3R_2 \\
R_4R_3R_2R_1R_4 & \rightarrow & R_3R_4R_3R_2R_1
\end{array}\right.$$

A remarkable feature of the systems S_n is that $S_n \subset S_{n+1}$. Thus the infinite set of rules $S_\infty = \bigcup_{n=1}^{\infty} S_n$ defines a canonical form of a permutation of arbitrary length. As for the surfaces fundamental groups, such a system must be compiled into three efficient algorithms:

- An algorithm of normalization, computing the relation $\xrightarrow[S_\infty]{*}$, using knowledge about the special form of the rules.
- An algorithm to perform the product of two permutations already in irreducible form, for which we know the localization of the possible reductions.
- A computation of P^{-1} from the normal form of P

6.5.2 Presentation with all transpositions

We put $T_{i,j}=(i\ j)$ with $1 \leq i < j \leq n$. These new generators are related to the previous ones by:

$$T_{i,j} = R_i R_{i+1} \cdots R_{j-2} R_{j-1} R_{j-2} R_{j-3} \cdots R_i$$

The definition of $\mathbf{S}_n$ is therefore:

$$\begin{cases} T_{i,j}^2 = 1 & \\ T_{i,j}T_{i,i+1} = T_{i,i+1}T_{i+1,j} & \\ (T_{i,i+1}T_{i+1,i+2})^3 = 1 & \\ (T_{i,i+1}T_{j,j+1})^2 = 1 & i+1<j \\ T_{i,i+1}T_{i+1,j}T_{i,i+1} = T_{i,j} & i+1<j \end{cases}$$

A possible completion is:

$$\begin{cases} T_{i,j}^{-1} \to T_{i,j} & \\ T_{i,j}T_{i,j} \to 1 & \\ T_{i,j}T_{k,l} \to T_{k,l}T_{i,j} & i\neq k, i\neq l, j\neq l \\ T_{i,j}T_{i,k} \to T_{i,k}T_{k,j} & i<k<j \\ T_{i,j}T_{k,j} \to T_{k,i}T_{i,j} & k<i<j \\ T_{i,j}T_{k,i} \to T_{k,i}T_{k,j} & k<i<j \end{cases}$$

Termination: Lex Ordering, with all the inverses greater than their corresponding generators and

$$T_{n-1,n} > T_{n-2,n} > \cdots > T_{1,n} > T_{n-2,n-1} > T_{n-3,n-1} > \cdots > T_{1,n-1} > \cdots > T_{1,2}.$$

The number of rules is $O(n^4)$, which is far from the upper bound we gave in chapter 4, here exponential: $(n-1)(n-2)n!$. Once more, we have $T_n \subset T_{n+1}$. Thus $T_\infty = \bigcup_{n=1}^{\infty} T_n$ reduces an arbitrary length permutation to a canonical form. Another prominent feature of this set is that it is a ***symmetrized*** set. Moreover, the rules enumerate all the quasi-commutativity laws between the transpositions, and these rules are sufficient to compute in $\mathbf{S}_n$.

Moreover, we can argue that the completion procedure is a good tool in the study of finitely presented groups. Unlike the Todd-Coxeter coset enumeration [Cox72] (cf. §3.2.2), it applies to infinite groups as well as to families of groups. In some cases where the procedure loops, the infinite set of rules may be finitely presented (see example $G3$ in §6.3). These facts follow from the observation that the rules usually present a simple structure, which in turn is the starting point for the design of word problem algorithms.

To conclude this last chapter, we may briefly compare the Todd-Coxeter coset enumeration [Tod36] and the Knuth-Bendix procedure for finite groups. In terms of Cayley diagrams, the coset enumeration computes a representation of the Cayley graph, while the completion, by a computation on its cycles, determines a unique path between two vertices. It is therefore obvious that, as quoted by R.H. Gilman

[Gil79], the coset enumeration is generally more efficient (cf. [Can73] for a detailed analysis of this algorithm). For the group E_6, M.F. Newman (private communication) reports that the Canberra implementation of coset enumeration produced a full coset table in less than three minutes, while we could not complete this group. The main advantages of completion technique in groups is its ability to handle parametrized classes, providing efficient word problems after an analysis of the canonical system. Moreover, the study of Coxeter groups has shown that infinite sets of rules could be described (see also [Ped84] as an example of using an infinite set of rules in solving the free word problem for the groupoid variety $(x.xy)x=y$). Therefore, these two algorithms appear to be complementary: one being well-suited for isolated groups, the other for parametrized families.

References

Ack54. Ackermann W., *Solvable cases of the decision problem*, North-Holland (1954).

Aho72. Aho A., Sethi R., Ullman J.D., *Code optimization and finite Church-Rosser theorems*, Proc. of Courant Comp. Sci. Symp. 5, Rustin R. (ed.), Prentice Hall (1972).

Ayo81. Ayoub C.W., *On constructing bases for ideals in polynomial rings over the integers*, Penn. State Univ., Univ. Park, Dept. Math., Rep. 8184 (1981).

Bac80. Bachmair L., Buchberger B., *A simplified proof of the characterization theorem for Gröbner-basis*, ACM SIGSAM Bull., 14,4, 29-34 (1980).

Bach85. Bachmair L., Plaisted D.A., *Associative path orderings*, Proc. of the First Int. Conf. on Rewriting Techniques and Applications (1985).

Bal79. Ballantyne A.M., Lankford D.S., *New decision algorithms for finitely presented commutative semigroups*, Math. and Comp. with Appl., 159-165 (1981).

Bal85. Balzac S.R., Davenport J.H., Gianni P., Jenks R.D., Miller V.S., Morrison S.C., Rothstein M., Sundaresan C.J., Sutor R.S., Trager B.M., *SCRATCHPAD II: An experimental computer algebra system*, Math. Sci. Dep., IBM Th.J. Watson Research Center, Yorktown Heights (1985).

Bar77. Barwise J. (ed.), *Handbook of Mathematical Logic*, North-Holland (1977).

Bar80. Barendregt H., *The Lambda-Calculus: Its Syntax and Semantics*, North-Holland (1980).

Bau81. Bauer G., *Zur darstellung von monoiden durch konfluente regelsysteme*, Dissertation, Universität Kaiserslautern (1981).

Bau84. Bauer G., Otto F., *Finite complete rewriting systems and the complexity of the word problem*, Acta Informatica 21, 521-540 (1984).

Ben85. Benanav D., Kapur D., Narendran P., *Complexity of matching problems*, Proc. of the First Int. Conf. on Rewriting Techniques and Applications (1985).

Ber78. Bergman G.M., *The diamond lemma for ring theory*, Adv. in Math., 29,2, 178-218 (1978).

Bir35. Birkhoff G., *On the structure of abstract algebras*, Proc5. Cambridge Phil. Soc. 31, 433-454 (1935).

Bir67a. Birkhoff G., *Lattice Theory*, AMS Colloq. Publ. 25, Providence, R.I. 3rd ed. (1967).

Bir71. Birkhoff G., Hall M.,Jr. (eds.), *Computers in Algebra and Number Theory*, SIAM-AMS Proc. 4, Providence (1971).

Bir67b. Biryukov A.P., *Some algorithmnic problems for finitely presented commutative semigroups*, Sib. Mat. Zh. 8,3, 525-534. (1967).

Ble84. Bledsoe W.W., Loveland D.W. (eds.), *Automated Theorem Proving: after 25 years*, Contemporary Mathematics, 29 (1984).

Blo84. Bloniarz P.A., Hunt H.B. III, Rosenkrantz D.J., *Algebraic structures with hard equivalence and minimization problem*, J. ACM, 31,4, 879-904 (1984).

Bok72. Bokut' L.A., *Unsolvability of the word problem, and subalgebras of finitely presented Lie algebras*, Izv. Akad. Nauk. S.S.S.R. Ser. Mat. 36-6, pp.1173-1219 (1972).

Bok80. Bokut' L.A., *Decision problems for ring theory*, in Word Problems II. Adjan S.I., Boone W., Higman G. (eds.), North-Holland (1980).

Boo59. Boone W., *The word problem*, Ann. of Math.2,70, 207-265 (1959).

Boo81. Book R.V., O'Dunlaing. C., *Testing for the Church-Rosser property*, Theo. Comp. Sci. 16, 223-229, North-Holland (1981).

Boo82a. Book R.V., *Confluent and other types of Thue systems*, J. ACM, 29,1, 171-182. (1982).

Boo82b. Book R.V., Jantzen M., Wrathall C., *Monadic Thue systems*, Theo. Comp. Sci. 19, 231-251, North-Holland (1982).

Boo83. Book R.V., *Decidable sentences of Church-Rosser congruences*, Theo. Comp. Sci. 24, 301-312, North-Holland (1983).

Boo85. Book R.V., *Thue systems as rewriting systems*, Proc. of the First Int. Conf. on Rewriting Techniques and Applications (1985).

Boo59. Boone W., *The word problem*, Ann. of Math. 2,70, 207-265 (1959).

Boo74. Boone W., Higman G., *An algebraic characterization of groups with solvable word problem*, J. Aust. Math. Soc. 2, 41-53. (1974).

Bou78. Bourbaki N., *Groupes et algèbres de Lie, Ch. 4,5,6*, Actu. Sci. Ind., Hermann, Paris. (1978).

Boy79. Boyer R.S., Moore J.S., *A Computational Logic*, Academic Press (1979).

Bra71. Bradley G.H., *Algorithms for Hermite and Smith normal matrices and linear diophantine equations*, Math. Comp., 25,116, 897-907 (1971).

Bro71. Brown W.S., *On Euclid's algorithm and the computation on polynomial greatest common divisor*, J. ACM, 18,4, 478-504 (1971).

Bru75. de Bruijn N.G., *Exact finite models for minimal propositional calculus over a finite alphabet*, T.H.-Report 75-WSK-02 (1975).

Bru80. de Bruijn N.G., *A survey of the project Automath*, In To H.B. Curry: Essays on Combinatory Logic, Lambda Calculus and Formalism, Seldin J.P., Hindley J.R. (eds.) (1980).

Büc79a. Bücken H., *Reduction systems and small cancellation theory*, Proc. 4th Workshop on Automated Deduction, 53-59 (1979).

Büc79b. Bücken H., *Reduktionssysteme und Wortproblem*, Rhein.-Westf. Tech. Hochschule, Aachen, Inst. für Inf., Rep. 3 (1979).

Buc65. Buchberger B., *Ein Algorithmus zum Auffinden der Basiselemente des Restklassenringes nach einem nulldimensionalen Polynomideal*, Dissertation, Innsbruck, Austria (1965).

Buc79. Buchberger B., *A criterion for detecting unnecessary reductions in the construction of Gröbner bases*, Proc. EUROSAM 79, LNCS 72, 3-21, Springer-Verlag (1979).

Buc81. Buchberger B., *H-bases and Gröbner-bases for polynomial ideals*, Johannes Kepler Universität, Austria (Feb. 1981).

Buc82a. Buchberger B., *Miscellaneous results on Gröbner-bases for polynomial ideals II*, Johannes Kepler Universität, Austria (June 1982).

Buc82b. Buchberger B., Collins, G.E., Loos R. (eds.), *Computer Algebra, Symbolic and Algebraic Computation*, Comp. Supp 4, Springer-Verlag (1982).

But84. Butler G., Lankford D., *Dickson's lemma, Hilbert's basis theorem, and applications to completion in commutative noetherian rings*, Tech. Rep., Dep. of Math. and Stat., Louisiana Tech. Univ. (1984).

Can73. Cannon J.J., Dimino L.A., Havas G., Watson J.M., *Implementation and Analysis of the Todd-Coxeter algorithm*, Math. Comp. 27, 463-490 (1973).

Car75. Cardoza E.W., *Computational complexity of the word problem for commutative semigroups*, M.I.T., Project MAC, TM67 (1975).

Car76. Cardoza E.W., Lipton R., Meyer A.R., *Exponential space complete problems for Petri nets and commutative semigroups*, Proc. Eighth Ann. ACM Symp. on Th. of Comp., 50-54 (1976).

Chu36. Church A., Rosser J.B., *Some properties of Conversion*, Trans. AMS 39, 472-482 (1936).

Chu41. Church A., *The Calculi of Lambda-Conversion*, Ann. of Math. Studies, 6, Princeton Univ. Press. (1941).

Cli67. Clifford A.H., Preston G.B., *The algebraic theory of semigroups, Vol II*, AMS Providence, Rhode Island (1967).

Coc76. Cochet Y., *Church-Rosser congruences on free semigroups*, Colloq. Math. Soc. Janos Bolyai: Algeb. theory of Semigroups. 20, 51-60 (1976).

Coh65. Cohn P.M., *Universal Algebra*, Harper and Row, New York (1965).

Coh73. Cohn P.M., *The word problem for free fields*, J. Symb. Log. 38, 309-314. Correction and Addendum: J. Symb. Log. 40, 69-74. (1973).

Col80. Collins G.E., *ALDES and SAC-2 now available*, SIGSAM Bull.14,2, 19 (1980).

Coo71. Cook S., *The complexity of theorem-proving procedures*, Proc. 3rd ACM Symp. on Theory of Comp. (May 1971).

Cos85a. Cosmadakis S.S., Kanellakis P.C., Spyratos N., *Partition semantics for relations*, Proc. of the Fourth ACM Symposium on Principles of Database Systems (1985).

Cos85b. Cosmadakis S.S., Kanellakis P.C., *Two applications of equational theories to database theory*, Proc. of the First Int. Conf. on Rewriting Techniques and Applications (1985).

Cox35. Coxeter H.S.M., *The complete enumeration of finite groups of the form* $R_i^2=(R_i R_j)^{k_{ij}}=1$, J. Lond. Math. Soc., 10, 21-25. (1935).

Cox72. Coxeter H.S.M., Moser W.O.J, *Generators and Relations for Discrete Groups*, Springer-Verlag (1972).

Cur58. Curry H.B., Feys R., *Combinatory Logic, Vol. I*, North-Holland (1958).

Dav58. Davis M., *Computability and Unsolvability*, McGraw-Hill, USA. (1958).

Dav65. Davis M., *The Undecidable, basic papers on undecidable propositions, unsolvable problems and computable functions*, Raven Press, Hewlett, NY (1965).

Dav73. Davis M., *Hilbert's tenth problem is unsolvable*, Am. Math. Mon 80,3, 233-269 (1973).

Dea56. Dean R.A., *Completely free lattices generated by partially ordered sets*, Trans. AMS. 83, 238-249 (1956).

Deh11. Dehn M., *Uber unendliche diskontinuierliche Gruppen*, Math. Ann. 71, 116-144 (1911).

Deh12. Dehn M., *Transformation der Kurve auf zweiseitigen Flache*, Math. Ann. 72, 413-420 (1912).

Der79a. Dershowitz N., *Orderings for term-rewriting systems*, Proc. 20th ACM Symp. FOCS, 123-131 (1979).

Der79b. Dershowitz N., Manna Z., *Proving termination with multiset orderings*, Comm. ACM 22, 465-476 (1979).

Dic13. Dickson L.E., *Finiteness of the odd perfect and primitive abundant numbers with n distinct prime factors*, Amer. J. Math. 35, 413-422 (1913).

DoL83. Do Long Van, *The word and conjugacy problems for a class of groups with nonhomogeneous conditions of small cancellation*, Arch. der Math. 41,6, 481-490, (1983).

Dwo83. Dwork C., Kanellakis P.C., Mitchell J.C., *On the sequential nature of Unification*, Tech. Rep. CS-83-26, Brown Univ., Providence (1983).

Emi63. Emilicev V.A., *On algorithmic decidability of certain mass problems in the theory of commutative semigroups*, Sibirsk. Mat. Z 4, 788-798 (1963).

Eva51a. Evans T. , *The word problem for abstract algebras*, J. London Math. Soc. 26, 64-71. (1951).

Eva51b. Evans T., *On multiplicative systems defined by generators and relations 1, Normal forms theorems*, Proc. Cambridge Phil. Soc. 47, 637-649 (1951).

Eva69. Evans T., *Finitely presented loops, lattices, etc. are hopfian*, J. London Math. Soc. 44, pp. 551-552 (1969).

Eva78a. Evans T., *Word problems*, Bull. AMS 84,5, 789-802 (1978).

Eva78b. Evans T., *An algebra has a solvable word problem if and only if it is embeddable in a finitely presented simple algebra*, Alg. Univ. 8, 197-204. (1978).

Eva78c. Evans T., *Some solvable word problems*, in Word Problems II, Adjan S.I., Boone W., Higman G. (eds.) North-Holland (1978).

Eva75. Evans T., Mandelberg K., Neff M.F., *Embedding algebras with solvable word problems in simple algebras. Some Boone-Higman type theorems*, Proc. Logic Colloq., Univ. of Bristol, 1973. North-Holland (1975).

Fag84a. Fages F., *Associative-commutative unification*, 7th Conf. Automated Deduction, LNCS 170, 194-208, Springer-Verlag (1984).

Fag84b. Fages F., *Le système KB : manuel de référence : présentation et bibliographie, mise en oeuvre*, GRECO de programmation (1984).

Fag83a. Fages F., *Formes canoniques dans les algèbres booléennes et application à la démonstration automatique en logique de premier ordre*, Thèse de 3ème cycle, Univ. de Paris VI (June 1983).

Fag83b. Fages F., Huet G., *Unification and matching in equational theories*, CAAP 83, LNCS 159, Springer-Verlag (1983).

Fat79. Fateman R.J., *MACSYMA's general simplifier : philosophy and operation*, Proc. MACSYMA Users' conference, Berkeley, Fateman R.J. (ed.), MIT Press, 563-582 (1979).

Fre80. Freeze R., *Free modular lattices*, Trans. AMS. 261,1, 81-91 (1980).

Gal85. Galligo A., *Some algorithmic questions on ideals of differential operators*, Proc. EUROCAL 85, to appear in LNCS, Springer-Verlag (1985).

Gen69. Gentzen G., *The collected papers of Gerhard Gentzen*, Szabo M.E., (ed.), North-Holland (1969).

Gil79. Gilman R.H., *Presentations of Groups and Monoids*, J. Alg. 57, 544-554. (1979).

Gil84. Gilman R.H., *Computations with rational subsets of confluent groups*, Proc. EUROSAM 84, LNCS 174, 207-212, Springer-Verlag (1984).

Giu85. Giusti M., *A note on the complexity of constructing standard bases*, Proc. EUROCAL 85, to appear in LNCS, Springer-Verlag (1985).

Goa80. Goad C., *Computational uses of the manipulation of formal proofs*, Ph.D. Thesis, Stanford Univ., Dep. Comp. Sci., Rep. Stand-CS-80-819 (1980).

Gol81. Goldfarb W.D., *The undecidability of the secondorder unification problem*, Theo. Comp. Sci., 13, 225-230, North-Holland (1981).

Gor79. Gordon M.J., Milner A.J., Wadsworth C.P., *Edinburgh LCF*, LNCS 78, Springer-Verlag (1979).

Grä78. Grätzer G., *General Lattice Theory*, Pure and Applied Mathematics, Academic Press, New-York (1978).

Grä79. Grätzer G., *Universal Algebra*, Springer-Verlag, New-York, 2nd ed. (1979).

Gre60a. Greendlinger M., *On Dehn's algorithm for the word problem*, Comm. Pure Appl. Math., 13, 67-83 (1960).

Gre60b. Greendlinger M., *On Dehn's algorithm for the word and conjugacy problems with applications*, Comm. Pure Appl. Math., 13, 641-677. (1960).

Gri71. Griesmer J.H., Jenks R.D., *SCRATCHPAD/1. An Interactive Facility for Symbolic Mathematics*, Proc. 2nd Sym. SAM, 42-58, Los Angeles (1971).

Gut84. Guttag J.V., Kapur D., Musser D.R., *Proc. of an NSF workshop on the rewrite rule laboratory*, Tech. Rep., General Electric CRD, Schenectady, NY 12345 (1984).

Hal58. Hall P., *Some word problems*, J. London Math. Soc. 33, 482-496 (1958).

Hal73. Hall M.Jr., *Computers in group theory*, Summer school on Group Theory and Computation, Galway (1973).

Hav77. Havas G., Newman M.F., *Application of computers to questions like those of Burnside*, in Burnside Groups Proc., Mennicke J.L. (ed.), Bielefeld, Germany (1977), LNM 806, 211-230, Springer-Verlag (1980).

Hav79. Havas G., Sterling L.S., *Integer Matrices and Abelian Groups*, Proc. EUROSAM 79, Marseille, LNCS 72, Springer-Verlag, 431-451 (1979).

Hen22. Hentzelt K., *Zur theorie der polynomideale und resultanten*, Math. Ann. 88, 53-79 (1922).

Her30. Herbrand J., *Recherches sur la théorie de la démonstration*, Thèse, Université de Paris (1930), In : Ecrits logiques de Jacques Herbrand, PUF Paris (1968).

Her26. Hermann G., *Die frage der endlich vielen schritte in der theorie der polynomideale*, Math. Ann. 95, 736-788 (1926).

Hil90. Hilbert D., *Uber die theorie der algebrischen formen*, Math. Ann. 36, 473-534 (1890).

Hin69a. Hindley R., *An abstract form of the Church-Rosser theorem I*, J. of Symb. Log. 34,4, 545-560 (1969).

Hin69b. Hindley R., *An abstract form of the Church-Rosser theorem II. Applications*, J. of Symb. Log. 39,1, 1-21 (1969).

Hir64. Hironaka H., *Resolution of singularities of an algebraic variety over a field of characteristic zero,I,II*, Annals of Math., 79, 109-326 (1964).

Hof82. Hoffmann C.M., *Group-theoretic algorithms and graph isomorphism*, LNCS 136, Springer-Verlag (1982).

Hop79. Hopcroft J.E. and Ullman J.D., *Introduction to automata theory, languages, and computation*, Addison-Wesley (1979).

Hor83. Horster P., *Reduktionssysteme, formale sprachen und automatentheorie*, Dissertation, TH Aachen (1983).

Hsi82. Hsiang J., *Topics in automated theorem proving and program generation*, Ph.D. Thesis, Univ. of Illinois at Urbana-Champaign. (Nov. 1982).

Hue75. Huet G., *A unification algorithm for typed λ-calculus*, Theo. Comp. Sci., 1, 27-57, North-Holland (1975).

Hue76. Huet G., *Résolution d'équations dans des langages d'ordre 1,2, ... ω*, Thèse d'Etat, Université de Paris VII (1976).

Hue78a. Huet G., *An algorithm to generate the basis of solutions to homogeneous linear Diophantine equations.*, Inf. Proc. Letters 7,3, 144-147 (1978).

Hue80a. Huet G., *Confluent reductions: abstract properties and applications to term rewriting systems*, J. ACM 27,4, 797-821 (1980).

Hue81. Huet G., *A complete proof of correctness of the Knuth-Bendix completion algorithm*, J. CSS 23,1, 11-21 (1981).

Hue78b. Huet G., Lankford D.S., *On the Uniform Halting Problem for Term Rewriting Systems*, Rapport Laboria 283, IRIA (1978).

Hue83. Huet G., Levy J.J., *Computations in non-ambiguous linear term rewriting systems*, Rapport de Recherche, INRIA (1983).

Hue80b. Huet G., Oppen D., *Equations and rewrite rules: a survey*, In Formal Languages: Perspectives and Open Problems, Book R.V. (ed.), Academic Press (1980).

Hul79. Hullot J.M., *Associative-commutative pattern matching*, Proc. 5th Int. Joint Conf. on Art. Int., Tokyo (1979).

Hul80a. Hullot J.M., *Compilation de formes canoniques dans les théories equationnelles*, Thèse de 3ème cycle, Université de Paris Sud (Nov. 1980).

Hul80b. Hullot J.M., *A catalogue of canonical term rewriting systems*, SRI Tech. Report (Mar. 1980).

Hun82. Hunt H.B. III, *Finite languages and the computational complexity of algebraic structures (an overview)*, State Univ. of New York at Albany, Comp. Sci. Dep., Tech. Rep., 82-14 (1982).

Jan81. Jantzen M., *On a special monoid with a single defining relation*, Theo. Comp. Sci., 16, 6-73, North-Holland (1981).

Jen74. Jenks R.D., *The SCRATCHPAD-Language*, ACM SIGSAM Bull. 8,2, 16-26 (1974).

Jen77. Jensen D, Pietrzykowski T., *Mecanizing ω-order type theory through unification*, Theo. Comp. Sci., 3, 123-171, North-Holland (1977).

Joh80. Johnson D.L., *Topics in the Theory of Group Presentations*, London Math. Soc. Lect. Note Series, 42, Cambridge Univ. Press, Cambridge (1980).

Jou83. Jouannaud J.P., *Confluent and coherent equational term rewriting systems, and applications to proofs in abstract data types*, CAAP83, Ausiello G., Protasi M. (eds.), LNCS 159 Springer-Verlag (1983).

Jou82. Jouannaud J.P., Reinig F., Lescanne J.P., *Recursive decomposition ordering*, in. Formal description of programming concepts 2, Bjorner D. (ed.), North-Holland (1982).

Jou84. Jouannaud J.P., Kirchner H., *Completion of a set of rules modulo a set of equations*, Proc. 11th ACM Conf. POPL (1984).

Kan84. Kandri-Rody A., Kapur D., *Algorithms for computing Gröbner bases of polynomial ideals over various euclidean rings*, Proc. EUROSAM 84, Fitch J. (ed.), LNCS 174, 195-206, Springer-Verlag (1984).

Kan85. Kandri-Rody A., Kapur D., Narendran P., *An ideal-theoretic approach to word problems and unification problems over finitely presented commutative algebras*, Proc. of the First Int. Conf. on Rewriting Techniques and Applications (1985).

Kap84. Kaplan S., *Conditional rewrite rules*, Theo. Comp. Sci., 33,2, 175-193, North-Holland (1984).

Kap85. Kapur D., Narendran P., *A finite Thue system with decidable word problem and without equivalent finite canonical system*, Theo. Comp. Sci., 35, 337-344, North-Holland (1985).

Kir84. Kirchner C., *A new equational unification method: a generalization of Martelli-Montanari algorithm*, Proc. 7th Conf. Automated Deduction, LNCS 170, Springer-Verlag (1984).

Kir85. Kirchner H., *Preuves par complétion dans les variétés d'algèbres*, Thèse d'Etat, CRIN, Univ. de Nancy I (1985).

Koz77a. Kozen D., *Complexity of finitely presented algebras*, Proc. of the Ninth annual ACM Symposium on Theory of Computing, ACM SIGACT (1977).

Koz77b. Kozen D., *Finitely presented algebras and the polynomial time hierarchy*, Technical Report 77-303, Dep. of Computer Science, Cornell Univ., Ithaca, New-York (1977).

Knu70. Knuth D.E., Bendix P., *Simple word problems in universal algebras*, In Computational Problems in Abstract Algebra, Leech J. (ed.), Pergamon Press, 263-297 (1970).

Kön03. König J., *Einleitung in die allgemeine theorie der algebraische grössen*, B.G. Teubner, Leipzig 1903 (1903).

Kru60. Kruskal J.B., *Well-quasi-orderings, the tree theorem, and Vazsonyi's conjecture*, Trans. AMS, 95, 210-225 (1960).

Lak70. Lakser H., *Normal and canonical representations in free products of lattices*, Can. J. Math. 22, 394-402 (1970).

Lal74. Lallement G., *On monoids presented by a single relation*, J. Alg. 32, 370-388 (1974).

Lan75. Lankford D.S., *Canonical algebraic simplification*, Univ. of Texas, Austin, Dept. Math Comp. Sci., Rep. ATP-32 (1975).

Lan79. Lankford D.S., *A unification algorithm for abelian group theory*, Tech. Rep. MTP-1, Math. Dep., Louisiana Tech. Univ. (Jan. 1979).

Lan77a. Lankford D.S., Ballantyne A.M., *Decision procedures for simple equational theories with commutative axioms: complete sets of commutative reductions*, Tech. Rep. ATP-35, Math. Dep., Univ. of Texas at Austin. (Mar. 1977).

Lan77b. Lankford D.S., Ballantyne A.M., *Decision procedures for simple equational theories with permutative axioms: complete sets of permutative reductions*, Tech. Rep. ATP-37, Math. Dep., Univ. of Texas at Austin. (Apr. 1977).

Lan77c. Lankford D.S., Ballantyne A.M., *Decision procedures for simple equational theories with associative-commutative axioms: complete sets of associative-commutative reductions*, Tech. Rep. ATP-39, Math. Dep., Univ. of Texas at Austin. (Aug. 1977).

Laz83. Lazard D., *Gröbner bases, Gaussian elimination and resolution of systems of algebraic equations*, Proc. EUROCAL 83, London, LNCS 162, 146-156, Springer-Verlag (1983).

LeC83. Le Chenadec P., *Formes canoniques dans les algèbres finiment présentées*, Thèse de 3ème cycle, Univ. d'Orsay (June 1983).

LeC84. Le Chenadec P., *Le système AL-ZEBRA pour algèbres finiment présentées*, Rapport GRECO (1984).

Leo77. Leon J.S., Sims C.C., *The existence and uniqueness of a simple group generated by {3-4}-transpositons*, Bull. AMS 83, 1039-1040 (1977).

Leo79. Leon J.S., Pless V., *CAMAC 1979*, Proc EUROSAM 79, Marseille, LNCS 72, Springer-Verlag, 249-257 (1979).

Les83. Lescanne P., *An introduction to REVE*, Rep. Greco 1.83, Université de Bordeaux I, Talence, France. (1983).

Lip64. Lipschutz S., *An extension of Greendlinger's results on the word problem*, Proc. AMS 15, 37-43 (1964).

Lip74. Lipshitz L., *The undecidability of the word problem for projective geometries and modular lattices*, Trans. AMS 193, 171-180 (1974).

Liv76. Livesey M., Siekmann J., *Unification of bags and sets*, Tech. Rep., Institut für Informatik, Univ. Karlsruhe (1976).

Liv79. Livesey M., Siekmann J., Szabo P., Unvericht E., *Unification problems for combinations of associativity, commutativity, distributivity and idempotence axioms*, 4th Workshop on Automated Deduction, Austin Texas, 161-167 (1979).

Llo83. Llopis de Trias R., *Canonical forms for residue classes of polynomial ideals and term rewriting systems*, Techn. Rep., Univ. Autonoma de Madrid, Div. de Mat. (1983).

Loo76. Loos R.G.K., *The algorithm description language ALDES (report)*, ACM SIGSAM Bull., 10, 14-38 (1976).

Lot83. Lothaire M., *Combinatorics on Words*, Encyc. of Math. and its Appl., Addison-Weslcy (1983)

Lyn66. Lyndon R.C., *On Dehn's algorithm*, Math. Ann. 166, 208-228 (1966).

Lyn77. Lyndon R.C., Schupp P.E., *Combinatorial Group Theory*, Springer-Verlag, Berlin. (1977).

Mac73. Macintyre A., *The word problem for division rings*, J. Symb. Log. 38, 428-436 (1973).

Mac77. Mathlab group, *MACSYMA Reference Manual, vers. 9*, M.I.T., Lab. Comp. Sci., Cambridge, Mass. (Dec. 1977).

Mag30. Magnus W., *Uber diskontinuierliche Gruppen mit einer definierenden Relation (Der Freiheitssatz)*, J. Reine Angew. Math. 163, 141-165 (1930).

Mag32. Magnus W., *Das identitatsproblem fur Gruppen mit einer definierenden Relation*, Math. Ann. 106, 295-307 (1932).

Mag66. Magnus W., Karrass A., Solitar D., *Combinatorial Group Theory*, Wiley, New York (1966).

Mak77. Makanin G.S., *The problem of the solvability of equations in a free semigroup*, Dokl. Akad. Nauk SSSR, 23,2 (1977), 287-290. Translated as Sov. Math. Dokl. 18,2, 330-334 (1977).

Mak82. Makanin G.S., *Equations in a free group*, Izv. Akad. Nauk SSSR, 46 (1982), 1199-1273. Translated as Math. USSR-Izv, 21,3, 483-546 (1983).

Mal58. Malcev A.I., *On homomorphisms of finite groups*, Ivano Gosudarstvennyi Pedagogicheskii Institut Uchenyi Zapiski 18, 49-60 (1958).

Man70. Manna Z., Ness S., *On the termination of Markov algorithms*, Proc. 3rd Hawaii Int. Conf. Sys. Sci. (1970).

Mar51. Markov A., *Impossibility of algorithms for recognizing some properties of associative systems*, Dokl. Akad. Nauk SSSR, 77, 953-956 (1951).

Mar82. Martelli A., Montanari U., *An efficient unification algorithm*, ACM TOPLAS, 4,2, 258-282 (1982).

Mat70. Matiyasevich Yu.V., *Enumerable sets are Diophantine*, Dokl. Akad. Nauk SSSR 191, 279-282 (1970). Translated as Sov. Math. Dokl. 11, 354-357 (1970).

May81. Mayr E.W., Meyer A.R., *The complexity of the word problems for commutative and polynomial ideals*, MIT Lab. Comp. Sci., Rep. LCS/TM-199 (1981).

Mét83. Métivier Y., *Systèmes de réécriture de termes et de mots*, Thèse de 3ème cycle, Université de Bordeaux I (May 1983).

Mil71. Miller C.F.III, *On group-theoretic decision problems and their classification*, Ann. of Math. Studies 68, Princeton Univ. Press, USA (1971).

Mil76. Miller C.F.III, *Decision problems in algebraic classes of groups*, In Word Problems. Boone W., Cannonito F., Lyndon R.C. (eds.), North-Holland, 507-523 (1976).

Möl84. Möller H.M., Mora F., *Upper and lower bounds for the degree of Gröbner bases*, Proc. EUROSAM 84, Fitch J. (ed.), LNCS 174, 172-183, Springer-Verlag (1984).

Mos66. Mostowski A.W., *On the decidability of some problems in special classes of groups*, Fund. Math. 59, 123-135 (1976).

Mus80. Musser D.L., *Abstract data type specification in the AFFIRM system*, IEEE Trans. Soft. Eng. (1980).

Nel79. Nelson C.G., Oppen D.C., *Simplification by cooperating decision procedures*, ACM TOPLAS 1,2, 245-257 (1979).

New42. Newman M.H.A., *On theories with a combinatorial definition of "equivalence"*, Ann. of Math. 43, 223-243 (1942).

New73. Newman M.F. (ed.), *Proc. 2nd Internat. Conf. on the Theory of Groups*, Canberra, Austral. Nat. Univ., LNM 372, Springer-Verlag (1974).

Nov55. Novikov P.S., *On the algorithmic unsolvability of the word problem in group theory*, Trudy Mat. Inst. Steklov, 44,143 (1955).

O'Du83. O'Dunlaing C., *Undecidable questions of Thue systems*, Theo. Comp. Sci., 23,3, 339-345, North-Holland (1983).

Oga78. Oganesjan G.U., *A class of semigroups with a solvable word problem*, Mat. Zam. 24,2, 259-265, 303 (1978). Translated as Math. Notes 23,5-6, 640-643 (1978).

Ott84a. Otto F., *Some undecidabiblity results for non-monadic Church-Rosser Thue systems*, Theo. Comp. Sci., 33, 261-278, North-Holland (1984).

Ott84b. Otto F., *Finite complete rewriting systems for the Jantzen monoid and the Greendlinger group*, Theo. Comp. Sci., 34, 249-260, North-Holland (1984).

Ott85. Otto F., *Deciding algebraic properties of monoids presented by finite Church-Rosser Thue systems*, Proc. of the First Int. Conf. on Rewriting Techniques and Applications (1985).

Pat78. Paterson M.S., Wegman M.N., *Linear unification*, J. CSS 16, 158-167 (1978).

Pau84. Paul E., *A new interpretation of the resolution principle*, Proc. 7th Conf. Automated Deduction, LNCS 170, 333-354, Springer-Verlag (1984).

Pau85. Paul E., *On solving the equality problem in theories defined by Horn clauses*, To appear in Proc. EUROCAL 85, Linz, LNCS, Springer-Verlag. (1985).

Péc81. Pécuchet J.P., *Equations avec constantes et algorithme de Makanin*, Thèse de 3ème cycle, Université de Rouen (Dec. 1981).

Ped84. Pedersen J.F., *Confluence methods and the word problem in universal algebra*, Ph.D. Thesis, Emory Univ., Dep. Math and Comp. Sci., Atlanta, Georgia, USA (1984).

Pet81. Peterson G.E., Stickel M.E., *Complete sets of reduction for equational theories with complete unification algorithms*, J. ACM 28,2, 233-264 (1981).

Pet82. Peterson G.E., Stickel M.E., *Complete systems of reductions using associative and/or commutative-unification*, SRI International (Oct. 1982).

Pig79. Pigozzi D., *Universal equational theories and varieties of algebras*, Ann. Math. Log. 17, 117-150 (1979).

Pla78a. Plaisted D., *Well-founded orderings for proving termination of systems of rewrite rules*, Report 78-932, Dept. Comp. Sci., Univ. of Illinois at , Urbana-Champaign. (Sept. 1978).

Pla78b. Plaisted D., *A Recursively defined ordering for proving termination of term rewriting systems*, Report 78-943, Dept. Comp. Sci., Univ. of Illinois at Urbana-Champaign. (Sept. 1978).

Plo72. Plotkin G., *Building-in equational theories*, Machine Intelligence 7, 73-90 (1972).

Pos47. Post E., *Recursive unsolvability of a problem of Thue*, J. Symb. Log. 12, 1-11 (1947).

Rab58. Rabin M.O., *Recursive unsolvability of group theoretic problems*, Ann. of Math., 67,1, 172-194 (1958).

Red63. Redei L., *Theory of finitely generated commutative semigroups*, Leipzig (1963).

Rob65. Robinson J.A., *A Machine-oriented logic based on the resolution principle*, J. ACM 12, 32-41 (1965).

Rob71. Robinson J.A., *Computational Logic : the Unification Computation* , Machine Intelligence 6, Meltzer B,, Michie D. (eds.), American Elsevier, New-York (1971).

Ros73. Rosen B.K., *Tree-manipulating systems and Church-Rosser theorems*, J. ACM 20, 160-187 (1973).

Sho67. Shoenfield J.R., *Mathematical Logic*, Addison-Wesley (1967).

Set74. Sethi R., *Testing for the Church-Rosser property*, J. ACM 21,4, 671-679 (1974).

Sie79. Siekmann J., *Unification of commutative terms*, Proc. EUROSAM 79, LNCS 72, 531-545, Springer-Verlag (1979).

Sie82. Siekmann J., Szabo P., *A nœtherian and confluent rewrite system for idempotent semigroups*, Semigroup Forum 25, 83-110 (1978).

Sie84. Siekmann J., *Universal unification*, Proc. 7th Conf. Automated Deduction, LNCS 170, 1-42, Springer-Verlag (1984).

Sir62. Sirsov A.I., *Some algorithmic problems for Lie algebras*, Sib. Math. J. 3,2, 292-296 (1962).

Sla74. Slagle R., *Automated theorem-proving for theories with simplifiers, commutativity and associativity*, J. ACM, 21,4, 622-642 (1974).

Smi66. Smith D.A., *A basis algorithm for finitely generated abelian groups*, Math. Algorithms, I,1 (Jan. 1966).

Sol69. Soldatova V.V., *Solution of the word problem for a certain class of groups*, Ivan. Gos. Ped. Inst. Ucen. Zap. 44, 17-25 (1969).

Sta75. Staples J., *Church-Rosser theorems for replacement systems*, In Algebra and Logic, Crossley J. (ed.), LNCS 450, Springer-Verlag, 291-307 (1975).

Sti81. Stickel M.E., *A complete unification algorithm for associative-commutative functions*, J. ACM 28,3, 423-434 (1981).

Sti84. Stickel M.E., *A case study of theorem proving by the Knuth-Bendix method. Discovering that $x^3=x$ implies ring commutativity*, Proc. 7th Conf. Automated Deduction, LNCS 170, 248-258, Springer-Verlag (1984).

Sti82. Stillwell J., *The word problem and the isomorphism problem for groups*, Bull. AMS (n.s.) 6,1, 33-56 (1982).

Szm54. Szmielew W., *Elementary properties of abelian groups*, Fund. Math. XLI, 203-271 (1954).

Tai74. Taiclin M.A., *The isomorphism problem for commutative semigroups*, Mat. Sb. (n.s.), 93,135, 103-128 (1974).

Tak69. Takeuchi K., *The word problem for free distributive lattices*, J. Math. Soc. Japan 21,4, 176-181 (1969).

Tal81. Talapov V.V., *On solvable Lie algebras with one defining relation*, Sib. Math. J. 22,4, 176-181 (1981).

Tal82. Talapov V.V., *Polynilpotent Lie algebras given by one defining relation*, Sib. Math. J. 23,5, 192-204 (1982).

Tar49a. Tartakovskii V.A., *The sieve method in group theory*, Mat. Sb. 25, 3-50 (1949).

Tar49b. Tartakovskii V.A., *Application of the sieve method to the solution of the word problem for certain types of groups*, Mat. Sb. 25, 251-274 (1949).

Tar49c. Tartakovskii V.A., *Solution of the word problem for groups with a k-reduced basis for k>6*, Izv. Akad. Nauk SSSR Ser Math, 13, 483-494 (1949).

Tit69. Tits J., *Le Problème des mots dans les groupes de Coxeter*, In: Sympos. Math. Rome 1967/68, 175-185. London Acad. Press (1969).

Tod36. Todd J.A., Coxeter H.S.M., *A practical method for enumerating cosets of a finite abstract group*, Proc. Edinb. Math. Soc. 2,5, 26-34 (1936).

Tse56. Tseiten G.S., *Associative calculus with unsolvable equivalence problem*, Dokl. Akad. Nauk SSSR, 107, 370-371 (1956).

Tur50. Turing A.M., *The word problem in semigroups with cancellation*, Ann. of Math. 2,52, 491-505 (1950).

Urs84. Ursic S., *A linear characterization of NP-complete problems*, Proc. 7th Conf. Automated Deduction, LNCS 170, 80-100, Springer-Verlag (1984).

Ust72. Ustjan A.E., *On the word problem for finitely presented semigroups*, Sib. Math. J. 13, 141-150 (1972).

Wal68. Waldhausen F., *The word problem in fundamental groups of sufficiently large irreducible 3-manifolds*, Ann. of Math. 2,88, 272-280 (1968).

Whi41. Whitman P., *Free lattices*, Ann. of Math. 2,42, 325-329 (1941).

Wic75. Wicks M.J., *Presentations of some classical groups*, Bull. Austr. Math. Soc. 13,1, 1-12 (1975).

Win84a. Winkler F., *The Church-Rosser property in computer algebra and special theorem proving: an investigation of critical-pair/completion algorithms*, Dissertation, Johannes Kepler Universität, Austria (May 1984).

Win84b. Winkler F., *On the complexity of the Gröbner basis algorithm over* $K[x,y,z]$, Proc. EUROSAM 84, Fitch J. (ed.), LNCS 174, 184-194, Springer-Verlag (1984).

Win83. Winkler F., Buchberger B., *A criterion for eliminating unnecessary reductions in the Knuth-Bendix algorithm*, Johannes Kepler Universität, Austria (Sep. 1983).

Yas70. Yasuhara A., *The solvability of the word problem for certain semigroups*, Proc. AMS 26, 645-650 (1970).